# MODELLING WITH PROJECTILES

## MATHEMATICS AND ITS APPLICATIONS

*Series Editor:* G. M. BELL, Professor of Mathematics,
King's College London (KQC), University of London

## STATISTICS AND OPERATIONAL RESEARCH

*Editor:* B. W. CONOLLY, Professor of Operational Research,
Queen Mary College, University of London

Mathematics and its applications are now awe-inspiring in their scope, variety and depth. Not only is there rapid growth in pure mathematics and its applications to the traditional fields of the physical sciences, engineering and statistics, but new fields of application are emerging in biology, ecology and social organization. The user of mathematics must assimilate subtle new techniques and also learn to handle the great power of the computer efficiently and economically.

The need for clear, concise and authoritative texts is thus greater than ever and our series will endeavour to supply this need. It aims to be comprehensive and yet flexible. Works surveying recent research will introduce new areas and up-to-date mathematical methods. Undergraduate texts on established topics will stimulate student interest by including applications relevant at the present day. The series will also include selected volumes of lecture notes which will enable certain important topics to be presented earlier than would otherwise be possible.

In all these ways it is hoped to render a valuable service to those who learn, teach, develop and use mathematics.

## Mathematics and its Applications

*Series Editor:* G. M. BELL, Professor of Mathematics, King's College London (KQC), University of London

*Series continued at back of book*

# MODELLING WITH PROJECTILES

DEREK HART, B.A.(Hons), D.A.S.E.
Head of Department of Mathematics

*and*

TONY CROFT, B.Sc., M.Phil.
Lecturer in Department of Mathematics
*both* of Crewe+Alsager College of Higher Education
Crewe, Cheshire

**ELLIS HORWOOD LIMITED**
Publishers · Chichester

Halsted Press: a division of
**JOHN WILEY & SONS**
New York · Chichester · Brisbane · Toronto

First published in 1988 by
**ELLIS HORWOOD LIMITED**
Market Cross House, Cooper Street,
Chichester, West Sussex, PO19 1EB, England
*The publisher's colophon is reproduced from James Gillison's drawing of the ancient Market Cross, Chichester.*


**Distributors:**

*Australia and New Zealand:*
JACARANDA WILEY LIMITED
GPO Box 859, Brisbane, Queensland 4001, Australia

*Canada:*
JOHN WILEY & SONS CANADA LIMITED
22 Worcester Road, Rexdale, Ontario, Canada

*Europe and Africa:*
JOHN WILEY & SONS LIMITED
Baffins Lane, Chichester, West Sussex, England

*North and South America and the rest of the world:*
Halsted Press: a division of
JOHN WILEY & SONS
605 Third Avenue, New York, NY 10158, USA

*South-East Asia*
JOHN WILEY & SONS (SEA) PTE LIMITED
37 Jalan Pemimpin # 05–04
Block B, Union Industrial Building, Singapore 2057

*Indian Subcontinent*
WILEY EASTERN LIMITED
4835/24 Ansari Road
Daryaganj, New Delhi 110002, India


© 1988 D. Hart and T. Croft/Ellis Horwood Limited

**British Library Cataloguing in Publication Data**
Hart, Derek, *1933–*
Modelling with projectiles.
1. Projectiles. Mathematical models
I. Title      II. Croft, Tony, *1957–*
531′.55

**Library of Congress Card No.** 88–761

ISBN 0–7458–0323–7 (Ellis Horwood Limited)
ISBN 0–470–21085–0 (Halsted Press)

Phototypeset in Times by Ellis Horwood Limited
Printed in Great Britain by Butler & Tanner, Frome, Somerset

# Table of contents

*To Gillian, Jan and Thomas Anthony*

# Preface

The aim of this book it to provide a readable, vivid and stimulating account of one of the most interesting, yet accessible, branches of theoretical mechanics. For countless years, Man has been faced with solving problems of projectile motion, often military, but also in sport and other fields. Many years of teaching mechanics to school and college students leave us in no doubt that projectile motion is one of the most inviting, absorbing and challenging fields of study. With even the minimum of basic mathematical techniques at his or her disposal the student can enjoy this rich store. Our approach throughout has been to provide a leisurely and readable tour, and to lead the student straight to the points of greatest interest, whilst remaining comprehensive. To this end, much necessary background material has been placed in appendices. This is in contrast with some books which deal first with vectors, differential equations, etc. Our aim is not to provide a textbook as such, but more a 'travel guide' to enable the motivated reader to experience the salient points. Where appropriate, throughout the book we have provided exercises which will enhance understanding and develop a feel for the subject, which we would strongly encourage the student to complete. We have made suggestions for the longer project-type of investigation. References after question numbers refer to articles noted in the annotated bibliography in Appendix IV.

Additionally we hope that sixth-form teachers and lecturers in Further and Higher Education establishments will find material suitable for class discussion and problem-solving sessions. As we have stated the study of projectile motion is one of the most rewarding and invigorating branches of mechanics. We hope that the reader will get as much enjoyment from this engaging field as we have done.

## ACKNOWLEDGEMENTS

Figs 1.1, 1.2, 1.3 and 1.4 are reproduced by permission of Gerald Duckworth.

We thank Prof. A. Tan and the Applied Probability Trust for permission to use Fig. 4.6.

We thank the Farming Press Ltd for permission to use the photograph which appears as Fig. 12.1.

# 1

# Introduction

From the earliest times, motion has intrigued the minds of men. Whether that motion was celestial or terrestrial, man has watched, studied, 'explained' and attempted its prediction. Well before the days of application, early Egyptians observed the motions of the sun, the moon, and the stars out of awe and curiosity and from their observations determined the patterns described by these heavenly bodies. Once these patterns were determined, over a period of many years, the Egyptians realised that certain terrestrial occurrences, such as floods, coincided with happenings in the sky. They learned to watch the sky carefully to observe these happenings, and in this way the early calendar was born.

The ancient Greeks spent much time trying to explain why bodies move as they do here on Earth. Aristotle (384–322 BC) argued that rest is the natural state of a body, so that any movement must be accompanied by some force that continues to act whilst the body remains in motion. This force would be proportional to the object's velocity. When a stone is dropped from the hand, the force in question is the weight of the stone, and heavier stones will fall faster than lighter ones. Aristotle knew, however, that when an object is dropped, it gains speed. Consequently, since its weight is constant, some other force must also be present to cause this gain in speed. He accounted for this by a rush of air exerting a force on the back of the body, thus increasing its velocity. Others had argued that a body moved more jubilantly as it approached its 'home'. Similarly, if an arrow is fired from a bow, the initial force of propulsion is provided by the tension in the bow, but thereafter a rush of air behind carries it to its target. It was this misguided principle, that a body's natural state is to be still and that a force is required to keep it in motion, that directed thinking in this field for many centuries.

While the intellectuals in these times were considering and debating causes and explanations of the movement of objects, the ordinary man was making use of such motion in his everyday life. Animals were hunted with a variety of moving objects from stones and spears, to arrows. Soldiers devised many cunning and clever devices to exploit the motion of a projectile: the finest of these early fighting machines were designed and built by the Greeks, and later by the Romans. They constructed various types of catapult to launch projectiles at their enemies. The onager (Fig. 1.1) consisted of a sling to hold a large stone. The arm was wound down almost to ground level by four soldiers, before being released, to hurl the stone forward.

The short-armed catapult (Fig. 1.2) had a cup-shaped depression rather than a sling in which to position the projectile. The end of the thick and heavy revolving arm projected through a skein of twisted sinew to which was attached a ratchet to enable the skein to be tensioned. An earlier and more efficient form had a slender, pliant arm and held a sling made of rope (Fig. 1.3).

The ballistics were extremely simple. The larger the device, the longer was its arm, the wider the arc traced out and therefore the greater the range. Only practical considerations limited the overall magnitude. The projectile was generally a stone, the weight of which varied according to the dimensions of the machine. To load and fire, the skein was torsioned, the sling was loaded, and the arm was racked back and held by a retaining catch. On operating the release gear, the arm flew forward hitting the crossbeam and so discharging the stone. Some of the more gigantic machines launched stones weighing up to 200 lb. The typical range for a 50 lb stone was about 500 yards.

Other catapulting devices relied not on twisted cords but on gravity, and worked by means of a counterweight. The trebuchet (Fig. 1.4) could hurl a 200–300 lb projectile over 600 yards. Some large examples had revolving arms of 50 feet in length and counterweights of 10 tons. Missiles other than stones were often used, such as dead horses, skulls and incendiary devices.

While projectiles were being launched the world over for one reason or another and despite the fact that Aristotle and others had wrestled with explanations of the causes of their motion, it was not until the seventeenth century that Galileo put forward scientific and mathematical arguments to predict their motion. He differed from his predecessors in that instead of being concerned with the reasons why projectiles moved, he concentrated upon describing such motions quantitatively, i.e. he introduced mathematical descriptions of physical phenomena, and in this sense he might be regarded as the father of mathematical modelling. Mathematical modelling consists of analysing a problem with the intention of extracting its important characteristics, and disregarding those which either do not or only peripherally affect the outcome. Aristotle had approached the problem of motion in terms of its origins, causes and effects. These concepts do not lead to straightforward quantification. Instead, Galileo disregarded characteristics such as these and concentrated upon measurable quantities such as weight, time, velocity and acceleration. From these he deduced formulae relating some to others. With the advent of formulae, mathematical arguments could be propounded.

Galileo considered the problem of a stone dropping from the hand. Importantly, he was not interested in why a stone falls, only that it does. Obviously the distance travelled by the stone increases as time elapses from the instant it is dropped. There is a relationship between the variables distance $d$, and time $t$, and by careful experiment and measurement Galileo obtained this in the form: distance is proportional to the square of the time elapsed, or

$$d \propto t^2$$

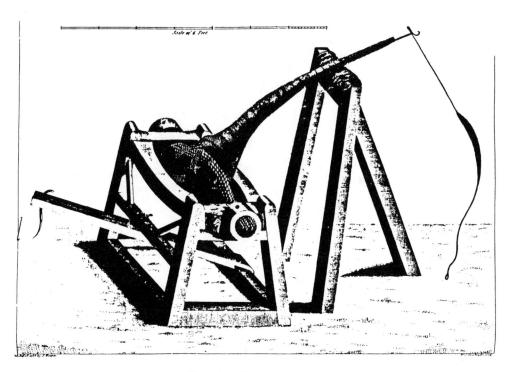

Fig. 1.1 — The onager.

Fig. 1.2 — A short-armed catapult.

Fig. 1.3 — A short-armed catapult.

This relationship is portrayed graphically in Fig. 1.5.

Galileo had greater insight than Aristotle. He realised that a body undisturbed by forces will continue in a straight line at a fixed speed. He was thinking of an object moving along a smooth horizontal surface and realised that if the motion was not impeded by air resistance or friction, the object would move indefinitely at a constant speed; in other words it possesses inertia. What Galileo was doing was mathematical modelling: he neglected certain effects (in this case friction and air resistance) and considered the outcome. It was this idealisation that paved the way for the tremendous strides in mechanics since his time. Of course, air resistance and friction are real phenomena, and in any particular situation they may have very important and not merely peripheral effects. But this causes no problem because when such effects are important they can be taken into account. This is all part of the modelling process: causes are neglected when their effects are small, but can be accounted for explicitly when required.

The modelling process can usefully be broken down into discrete stages as Fig. 1.6 shows. If we find that the solution to the problem is meaningless because we have neglected friction and air resistance, we can account for these where necessary. This is the 'refinement of the model' stage.

In the following chapters a variety of modelling examples are given and it will become apparent how the above stages are treated in order to solve the problems posed.

Galileo realised that if a force is applied to an object the object will gain or lose velocity, i.e. it will accelerate or decelerate. Conversely, if a body accelerates, some force must exist to produce this acceleration. An object falling to Earth, accelerating, must be acted upon by some force. In this case the Earth pulls the object down:

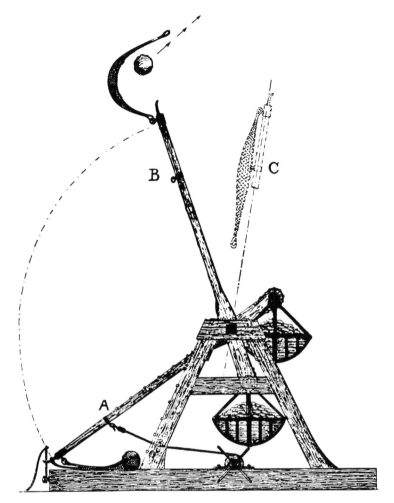

Fig. 1.4 — The trebuchet.

the force is the force of gravity. Galileo did not waste time speculating about gravity but instead concentrated upon quantitative facts about falling bodies. He discovered that in the absence of air resistance, all bodies falling to the surface of the Earth have the same constant acceleration, i.e. they gain velocity at the same rate, $9.8\,\mathrm{ms}^{-1}$ each second. If a body is dropped with zero speed, then after one second its speed is 9.8 $\mathrm{ms}^{-1}$, after two seconds, $19.6\,\mathrm{ms}^{-1}$ and so on. Galileo deduced that the speed, $V$, of the body was proportional to the time elapsed (Fig. 1.7).

From the two expressions relating distance and time, and speed and time, Galileo deduced that the time required for an object to fall a given distance is independent of its mass. Hence all bodies take the same time to fall a given distance. This is the lesson he supposedly learned by dropping objects from the Tower of Pisa.

He went on to deduce another law: if one body carries another, the carried body shares the motion of the first. This explains why, when we jump up from the Earth,

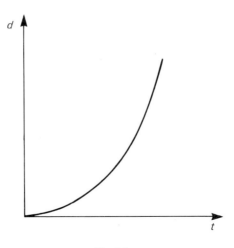

Fig. 1.5

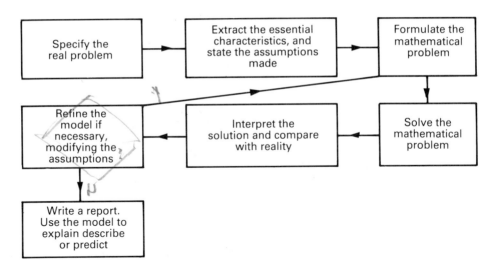

Fig. 1.6 — Mathematical modelling.

we are not left behind by its rotation: the Earth's surface is spinning at some $470\ \text{ms}^{-1}$ but we already possess its motion. While studying the motion of projectiles he observed that an object's motion can result from two independent simultaneous motions. Consider a projectile fired horizontally from the top of a cliff. Its motion can be divided into a horizontal motion and a vertical motion. The horizontal motion carries the projectile forward at the constant speed of projection. The vertical motion is solely due to gravity and is accelerated. The combination of the two causes the object to travel downwards on a parabolic path (Fig. 1.8).

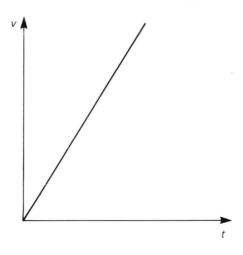

Fig. 1.7

However, the horizontal and vertical motions are independent: one can be varied without affecting the other. A greater initial horizontal speed of projection would have no effect on the time taken to reach the ground since this time depends only on the vertical distance the projectile has to travel. In other words, the time taken to travel paths P and Q in Fig. 1.9 is the same.

Galileo applied this principle of simultaneous motions to the motion of a cannon ball and proved that the path in this case too is parabolic, and that the greatest range is achieved by firing at an angle of 45°. All these results are expounded in great detail in his work 'Discourses and Mathematical Demonstrations Concering Two New Sciences', a masterpiece on which he worked for over 30 years.

While Galileo, through careful reasoning, grasped the relationship between force and motion, the truly quantitative appreciation came several decades later from Sir Isaac Newton (1642–1727). Newton was aware of Galileo's first law, that bodies should move in straight lines unless disturbed by forces. He was also aware of the work of Kepler (1571–1630) which stated that the planets move in elliptical orbits around the Sun. He concluded that some force must therefore be acting to deflect them from their straightline paths, and that this same force accounted for the fact that a body released from the hand falls to the Earth. Since both the Earth and the Sun attract bodies, the idea of unifying both actions under one theory was proposed and Newton was able to show that the very same mathematical formula describes the action of the Sun on the planets as well as the action of the Earth on objects near it. He concluded that the same force operates in both cases, and the Universal Theory of Gravitation was born. It was in the *Principia* that Newton described this theory and also stated formally the three laws of motion which now bear his name, even though the first two were known to Galileo. Readers are, no doubt, familiar with these laws, but for completeness we state the first two here as specifically applied to projectile motion.

(I)   If no forces are acting on a projectile which is at rest, then it will always remain at

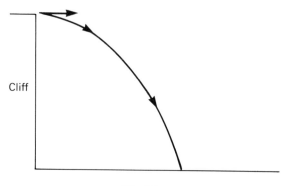

Fig. 1.8

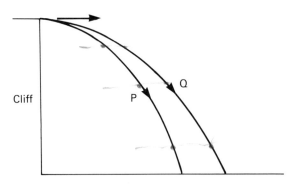

Fig. 1.9

rest. Conversely, a projectile at rest experiences no force, or, as is more likely, the resultant force is zero. (When this resultant force is zero we say there is no net force.) If no forces are acting on a projectile moving at a constant speed in a fixed direction, then it will always travel at that speed in that direction. Conversely, a projectile moving with a constant speed in a fixed direction experiences no net force. Finally, if a projectile's speed or direction changes, a force is required to produce this change. For example, we know from experience that a ball thrown from the hand will move in a roughly parabolic path. Since the direction of motion, and indeed the speed, changes, we conclude from this law that a force must be acting on the ball.

(II) If a force acts on a projectile, then the projectile accelerates in the direction of the force. The magnitude of this acceleration is equal to that of the force divided by the projectile's mass.

We will additionally make use of Newton's Universal Law of Gravitation which for our purposes, may be stated as follows: a projectile of mass $m$ experiences a force, $F$, which is known as weight, due to the presence of the Earth, and which is directed towards the centre of the Earth. The magnitude of this force is proportional to the mass of the projectile and inversely proportional to the square of the distance $r$, from the centre of the Earth. This can be written in vector notation as

$$\mathbf{F} = -\frac{km}{r^2}\,\hat{\mathbf{r}}$$

*outward*

where $\hat{\mathbf{r}}$ is a unit vector along the Earth's radius, through the projectile, and $k/r^2$ is the acceleration due to gravity. This quantity is conventionally denoted by $g$. Since $g$ depends upon $r$, the projectile's acceleration will change as its distance from the centre of the Earth changes. In the models considered herein, all motion will take place close to the surface of the Earth, so that we can take $g = k/R^2 = $ constant where $R$ is the mean radius of the Earth. This yields a value for $g$ of 9.8 ms$^{-2}$. Readers can easily verify that the value of $g$ at a point 100 km above the surface of the Earth is 9.52 ms$^{-2}$ — only a 3% change! The constant $g$ approximation is therefore justified for 'near-Earth' motions.

It should be noted that since the radius of the Earth changes from location to location, and also because the rotation of the Earth provides an additional 'gravitational-type' effect, $g$ is not in fact a constant, even near the surface of the Earth, and on occasions it will be necessary to take account of this variation explicitly (see Chapter 6).

Much of the modelling work with projectiles since the time of Newton has been based upon his ideas. Not only were his laws of motion vital, he did invaluable work on resisted motion, the study of which is important in many fields, especially in ballistics. A cannonball projected from the Earth at 800 ms$^{-1}$, at 30° to the horizontal, would have a horizontal range of 56.6 km, if there were no atmosphere, but in still-air it can hardly achieve half this distance. The effects of air resistance are extremely complicated and even today are not completely understood. However, at moderate speeds, an approximate experimental law is available which states that the resistive force, i.e. the drag, is proportional to the square of the speed. This law and its variants will be discussed in Chapter 5. The effect of resistance is to cause a retardation of the projectile, and over the years there have been many attempts to measure this. A notable method was devised by the Rev. Francis Bashforth in 1864 at the Royal Military College of Science. A series of screens of copper wire were placed in the path of the projectile, as illustrated in Fig. 1.10. These screens were connected to apparatus which measured the electric current allowed to flow in them. As the projectile passed through, the circuits were broken in succession, allowing the times of passage to be recorded. From these times the average velocity over the measured interval could be calculated and the resistance deduced.

Now more than ever before, modern military requirements necessitate the precise study of ballistics, and many previously disregarded 'peripheral' causes must be built into ever more complex mathematical models. With the advent of the space age the study of the motion of missiles and re-entry vehicles has attracted a great deal of attention. The mathematics involved in such studies is complicated but has its beginnings in the theory of projectiles. There are still down-to-earth problems. Numerous applications arise in sport, from the calculation of the range of a shot put

Fig. 1.10 — Bashforth's method for measuring air resistance.

to the modelling of the throw of a javelin to optimise design features (Chapters 6 and 10). In Chapter 9 we will use the theory of projectile motion to study the problem of grit being thrown up from a newly surfaced road. Other problems exist which have only intrinsic value such as the Percy Grainger problem in Chapter 8. The following chapters will give the necessary background, and then a variety of modelling problems are posed, some are solved and some are left for the reader to finish.

# 2

# Preliminary theory of projectile motion

## 2.1 THE BASIC EQUATIONS

There are several ways to represent the motion of a projectile. One of them is to consider the displacement vector, **r**, of the projectile relative to some fixed point O. The fixed point, O, is called the origin, and in the majority of applications it is the point from which the projectile was projected, as shown in Fig. 2.1.

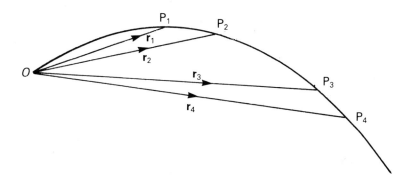

Fig. 2.1.

$P_1, P_2, \ldots$ are the successive positions of the projectile and $\mathbf{r}_1, \mathbf{r}_2, \ldots$ are the corresponding position or displacement vectors. In most applications the successive position vectors lie in a vertical plane and so it is useful to introduce a set of Cartesian axes with O as origin. Since the 'driving force' of a projectile is its weight, which can frequently be represented by a constant vertical vector, it is usual, but not essential, that the coordinate system $Oxy$ is such that $Ox$ is horizontal and $Oy$ is vertical, as illustrated in Fig. 2.2.

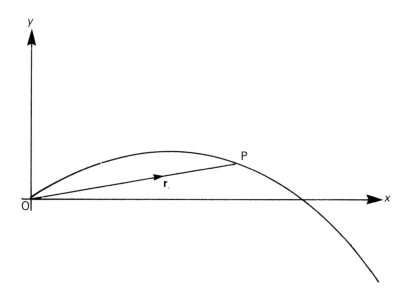

Fig. 2.2.

The locus of the points $O, P_1, P_2, \ldots$ is called the *trajectory*, and it is useful to have its Cartesian equation. We obtain this from the equation of motion of the projectile later in this chapter. Let $\mathbf{i}$, $\mathbf{j}$ be unit vectors in the directions $Ox, Oy$ respectively (See Appendix I). In the absence of air resistance the only force on the projectile is its weight, $mg$, where $m$ is the mass of the projectile and $g$ is the acceleration due to gravity. Using Newton's second law, the equation of motion can be written:

$$m\ddot{\mathbf{r}} = -mg\mathbf{j}$$

$$\Rightarrow \ddot{\mathbf{r}} = \begin{pmatrix} 0 \\ -g \end{pmatrix} \tag{2.1}$$

Before we can integrate this differential equation we need to know some initial conditions. Let us suppose that the projectile had initial velocity

$$\mathbf{V}_0 = \begin{pmatrix} u_0 \\ w_0 \end{pmatrix}$$

and initial position vector

$$\mathbf{r}_0 = \begin{pmatrix} 0 \\ 0 \end{pmatrix}$$

We thus have from (2.1), by integration,

$$\dot{\mathbf{r}} = \begin{pmatrix} u_0 \\ w_0 - gt \end{pmatrix} \tag{2.2}$$

Further integration yields

$$\mathbf{r} = \begin{pmatrix} u_0 t \\ w_0 t - \dfrac{gt^2}{2} \end{pmatrix} \tag{2.3}$$

Equations (2.2) and (2.3) can be used to obtain all the standard projectile results.

## 2.2  SOME STANDARD RESULTS

Although we have given the initial velocity in the form $(u_0 \ w_0)^{\mathrm{T}}$ it is often more convenient to write $u_0 = V_0 \cos\theta$ and $w_0 = V_0 \sin\theta$ where $V_0$ is the initial speed of projection at angle $\theta$ to the horizontal, as illustrated in Fig. 2.3.

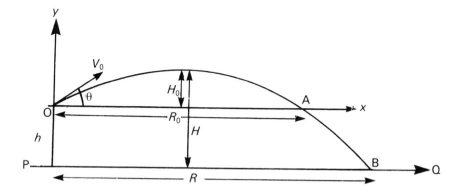

Fig. 2.3.

We should emphasise at this stage that the horizontal axis $Ox$ is a convenient mathematical fiction. It may or may not (usually not) correspond to a physical line as drawn on horizontal ground. Ground level is the straight line PQ, a vertical distance $h$ below $Ox$. We shall now calculate some useful formulae from equations (2.2) and

(2.3). $H_0$ is the maximum height of the projectile above $Ox$. $H$ is the maximum height above ground level. Clearly $H = H_0 + h$. $R_0 = OA$ is the horizontal displacement along $Ox$. $R$ is the horizontal displacement at B (i.e. the range). Let $t_H$ be the time taken for the projectile to travel from O to its maximum height. Let $t_A$ and $t_B$ be respectively the times taken to reach A and B. When the projectile is at its maximum height it is instantaneously travelling horizontally, i.e. the gradient of its path is zero and its vertical velocity is zero. It follows from (2.2) that

$$w_0 - gt_H = 0$$

so that

$$t_H = \frac{w_0}{g} = \frac{V_0\sin\theta}{g} \tag{2.4}$$

Hence $H_0$ is obtained by substituting (2.4) in (2.3):

$$H_0 = w_0 t_H - \frac{gt_H^2}{2}$$

$$= \frac{w_0 w_0}{g} - \frac{gw_0^2}{2g^2}$$

$$H_0 = \frac{w_0^2}{2g} = V_0^2 \frac{\sin^2\theta}{2g} \qquad \text{or} \qquad 0 = W_0^2 - 2g\,H_0 \tag{2.5}$$

And so

$$H = \frac{V_0^2 \sin^2\theta}{2g} + h$$

In a similar way we can find $t_A$ by putting $y = 0$ in (2.3) to obtain

$$w_0 t_A - \frac{gt_A^2}{2} = 0$$

i.e.

$$t_A = \frac{2w_0}{g}$$

$R_0$ is also found from (2.3)

$$R_0 = u_0 t_A$$

$$= \frac{2u_0 w_0}{g}$$

$$R_0 = \frac{2V_0^2 \sin\theta\cos\theta}{g} = \frac{V_0^2 \sin 2\theta}{g} \tag{2.6}$$

If we put $y = -h$ in (2.3) and proceed as above we obtain an expression for $R$:

$$-h = w_0 t_B - \frac{g t_B^2}{2}$$

i.e.

$$t_B^2 - \frac{2w_0 t_B}{g} - \frac{2h}{g} = 0$$

$t_B$ is the positive root of this quadratic equation, i.e.

$$t_B = \frac{1}{g} \left( w_0 + \sqrt{w_0^2 + 2hg} \right)$$

Now $x = u_0 t$

so that

$$R = u_0 t_B$$

$$\Rightarrow R = \frac{1}{g} \left( u_0 w_0 + \sqrt{u_0^2 w_0^2 + 2hg u_0^2} \right) \tag{2.7}$$

Note that the range depends crucially upon the height of release and that if $h = 0$ this formula reduces to that for $R_0$.

We now derive the Cartesian equation of the trajectory of the projectile. Equation (2.3) can be rewritten as

$$x = u_0 t$$

$$y = w_0 t - \frac{gt^2}{2}$$

The first of these equations gives $t = x/u_0$, and putting this in the second one we obtain

$$y = w_0 \frac{x}{u_0} - \frac{gx^2}{2u_0^2} \tag{2.8}$$

$$= x\tan\theta - \frac{gx^2}{2V_0^2 \cos^2\theta}$$

$$y = x\tan\theta - \frac{gx^2(1 + \tan^2\theta)}{2V_0^2} \tag{2.9}$$

This quadratic function represents a parabola. We deduce that the path of the projectile is a parabola which passes through the point of projection.

Very often in applications of the theory of projectiles we are concerned with finding maximum values of $R$ and $R_0$ for a given value of $V_0$. Applications of this kind are best considered by using the theory of the Enveloping Parabola which is dealt with in the next chapter.

## 2.3 WORKED EXAMPLES

1. A bullet is fired horizontally and strikes a target 100 m away from the point of projection at a distance 10 cm below the level of projection. What was the initial velocity of the bullet?

Solution:

The equation of the trajectory shown in Fig. 2.4 is obtained from (2.9) with $\theta = 0$:

$$y = \frac{-gx^2}{2V_0^2}$$

Clearly, when $x = 100$ m, $y = -0.10$ m. Taking $g = 10$ ms$^{-2}$, and solving this equation for $V_0$, gives $V_0 = 707$ ms$^{-1}$.

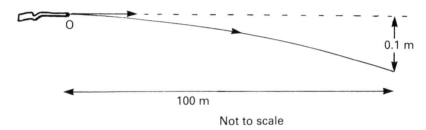

100 m

Not to scale

Fig. 2.4 — Trajectory of the bullet.

2. When a try has been scored in rugby, the player taking the kick at goal ('converting the try') is allowed to move as far downfield as he wishes from the point where the try was scored to place the ball for the kick, as shown in Fig. 2.5.

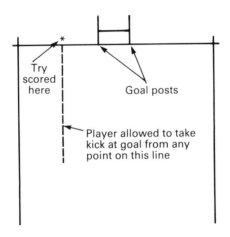

Fig. 2.5.

In order to convert the try, the ball must be kicked over a bar of height $h$. Suppose that a particular try is scored directly under the posts. It is to be converted from a point on the ground a horizontal distance $d$ away from the bar. Obtain an expression for the speed of projection in terms of the angle of projection in order that the ball just clears the bar. Hence determine the angle for which minimum speed of projection is possible.

Solution:

The equation of the trajectory shown in Fig. 2.6 is (2.9):

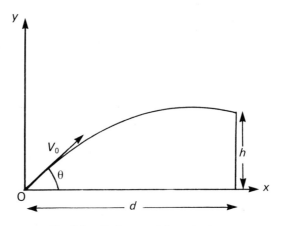

Fig. 2.6 — Trajectory of the rugby ball.

$$y = x\tan\theta - \frac{gx^2(1 + tan^2\theta)}{2V_0^2}$$

When $x = d$, $y = h$ so that

$$h = d\tan\theta - \frac{gd^2(1 + tan^2\theta)}{2V_0^2}$$

The required expression is therefore

$$V_0 = \sqrt{\frac{gd^2(1 + tan^2\theta)}{2(d\tan\theta - h)}}$$

Differentiating the expression for $V_0^2$ with respect to $\theta$ and setting this equal to 0 gives a value of $\theta$ of $\tan^{-1}(h + \sqrt{(h^2 + d^2)})/d$. Kicking the ball at this angle will require the least possible speed. An alternative approach is discussed in section 3.3.

3. A mathematics student is having a snowball fight with a friend. He knows that a snowball can be thrown at two different angles of projection, but with the same speed, and still reach the same target. However, the two times of flight will differ. The friend has developed the knack of catching the snowball and throwing it back. In order to obtain the upper hand, the mathematics student decides to launch two snowballs, at different times, one at a higher trajectory than the other. This will

act as a decoy and whilst the friend is watching and waiting to catch it, the second ball arrives and hits him simultaneously. If the friends are 25 m apart and they throw at 20 ms$^{-1}$, what are the two angles of projection, and what is the time delay between throwing the snowballs?

Solution:

Put $x = 25$ m and $y = 0$ in the equation of the trajectory. This provides a quadratic equation, the roots of which are the tangents of the two angles of projection, 70.7° and 19.3°. Recall that $t_A = 2V_0\sin\theta/g$ so that the corresponding times of flight are 3.8 and 1.3 s respectively. Since the first is to be a decoy, the second should be thrown 2.5 s after the first in order to strike the friend with surprise!

## EXERCISES

1.  A ball is thrown horizontally at 20 ms$^{-1}$ and hits a target 10 m away from the point of projection. Recall that Galileo's principle of simultaneous motions allows us to consider the horizontal and vertical motions separately. On a piece of graph paper, plot the horizontal and vertical positions of the ball at times 0.1, 0.2, . . . seconds, taking the point of projection as the origin of coordinates. Continue plotting the position of the ball until it reaches its target. Estimate the distance the ball drops before hitting the target (take $g = 10$ ms$^{-2}$). Now verify your result by calculation.

2.  A basketball is thrown in a gymnasium at a height of 2.1 m, at a speed of 8 ms$^{-1}$, and at an angle of 64° to the horizontal. Are the ceiling lightfittings at a height of 4.8 m likely to interfere with such a throw?

3.  (French) When firing an arrow from a bow, the angle of projection is generally quite small, so that the small-angle approximations can be used to simplify some of the formulae derived in this chapter. Recall that for small $\theta$ measured in radians, $\sin\theta \simeq \theta$, $\tan\theta \simeq \theta$, and $\cos\theta \simeq 1$. Verify that for small $\theta$, in radians, the range can be expressed as $R_0 \simeq 2\theta V_0^2/g$.

    In archery, the range is known so we can use this formula to find an approximation to the angle of projection, i.e. $\theta \simeq gR_0/(2V_0^2)$. Typical values of $R_0$ and $V_0$ are 60 m and 60 ms$^{-1}$ resp. Obtain $\theta$ for these and other typical values and compare your answers with the exact ones obtained from the equations derived earlier in the chapter.

    Using the same idea, can you find an approximate formula for the time of flight?

4.  A train is travelling at 20 kmh$^{-1}$ and a man on the roof throws a ball vertically upwards at 20 ms$^{-1}$, relative to his frame of reference.

    Describe the path of the ball as observed by the man. Which of Galileo's laws have you used? Now describe the path as observed by someone standing alongside the railway line. Where does the thrower appear to be in relation to the ball?

5.  (Hughes) A service shot at lawn tennis is typically taken from a point 12 m from the net, and the ball is struck high above the player's head, say at a height of 2.5 m in a direction perpendicular to the net. The ball must pass over the net which is of

height about 1 m and must land in the opposite service court which extends a distance of 7 m. Write down the constraints on $y$ needed to satisfy these two conditions and show that the possible angles of projection depend upon the speed of release $V$. Show that for a successful shot, $V$ must be less than some maximum and find the range of angles of projection for a number of values of $V$ less than this maximum.

Extend this exercise to the case of a more general service shot.

6. A hunter aims his rifle horizontally at a monkey hanging on the branch of a tree. At the instant the bullet is fired from the rifle, the monkey releases its grip on the branch and starts to fall to the ground. Investigate whether or not the hunter hits his target.

He now wishes to shoot another monkey which is near the top of a tree. If he knows the monkey will release its grip at the moment of firing, investigate how the hunter should aim his rifle to be sure of a hit.

7. (Beevers) A fireman is directing water into a window at a height $H$ above the ground. Because of the heat, he wishes to stand as far back from the wall as possible. The speed of projection is $V_0$. Find the maximum distance the fireman can stand from the foot of the wall assuming that the water moves on a parabolic path like a projectile.

8. (Porter) (a) What is the smallest angle of projection at which a player can score at basketball? (b) What is the minimum initial velocity?

9. Fig. 2.7 shows a diagram which appears in many school physics textbooks. It

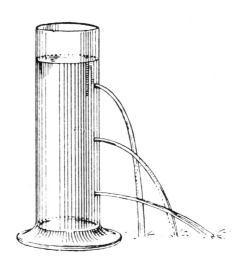

Fig. 2.7.

purports to demonstrate that water pressure depends upon depth. Given that the velocity of the water emerging from the holes is proportional to the square root of the depth, show that this commonly occurring diagram is incorrect.

# 3

# The enveloping parabola

## 3.1 DERIVATION OF THE ENVELOPING PARABOLA

It is intuitively obvious that if a projectile is launched from an origin O with speed $V_0$ there are certain regions of the $xy$ plane within its range, and there are other regions which are not. In the absence of air resistance the dividing curve between these regions has become variously known as the Bounding Parabola, the Enveloping Parabola, or the Parabola of Safety. In this chapter we develop the theory of this special curve and show how it can be used to solve many maximum-range-type problems.

Consider the family of curves obtained when we plot the trajectories of a projectile, launched with a given, fixed speed $V_0$, at various angles of projection $\theta$. A number of such curves are shown in Fig. 3.1.

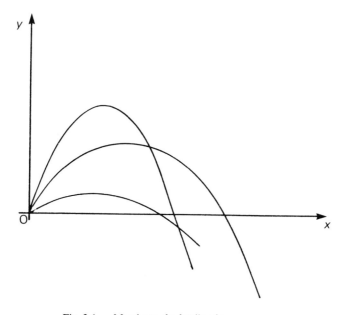

Fig. 3.1 — Members of a family of curves.

If we add more and more members of the family, it soon becomes apparent that there is a curve, which is not a trajectory, which touches all the trajectories. This is the dashed curve in Fig. 3.2.

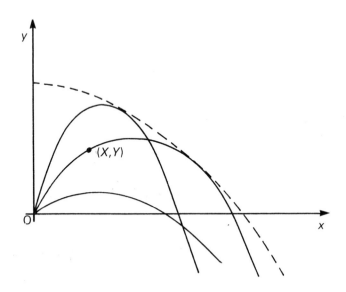

Fig. 3.2 — The envelope.

Since all the trajectories lie under this curve, it is the dividing curve between so-called 'accessible' regions and 'inaccessible' regions of the $xy$ plane. We now develop its equation in two ways.

Recall from (2.9) that the equation of the trajectory is given by

$$y = x\tan\theta - \frac{gx^2(1+\tan^2\theta)}{2V_0^2} \tag{3.1}$$

which we can write in the form

$$x\tan\theta - \frac{gx^2(1+\tan^2\theta)}{2V_0^2} - y = 0 \tag{3.2}$$

A point $(X,Y)$ is accessible if some trajectory passes through it as shown in Fig. 3.2. In this case the corresponding angle of projection is found by solving

$$X\tan\theta - \frac{gX^2(1+\tan^2\theta)}{2V_0^2} - Y = 0 \tag{3.3}$$

This is a quadratic equation for $\tan\theta$ which can be written in the form

$$\tan^2\theta - \frac{2V_0^2\tan\theta}{gX} + \left(1 + \frac{2YV_0^2}{gX^2}\right) = 0 \tag{3.4}$$

It is well known that a quadratic equation $ax^2 + bx + c = 0$ possesses

> real roots if $b^2 - 4ac > 0$
>
> equal roots if $b^2 - 4ac = 0$

and

> complex roots if $b^2 - 4ac < 0$

In the last case there are clearly no real values of $\tan\theta$ for which a trajectory passes through $(X,Y)$, and thus $(X,Y)$ is in the inaccessible region of the $xy$ plane. On the other hand, if there are two real distinct roots then $(X,Y)$ can be reached along two different trajectories. If only equal roots exist then $(X,Y)$ is reached by only one trajectory. This case divides the inaccessible from the accessible regions, so it is precisely this case which represents the enveloping parabola. Applying the condition for equal roots we find

$$\frac{4a^2}{X^2} = 4\left(1 + \frac{2Ya}{X^2}\right)$$

where $a = V_0^2/g$.

$$\Rightarrow a^2 = X^2 + 2Ya$$

We conclude that the equation of the enveloping parabola is

$$2ay = a^2 - x^2 \tag{3.5}$$

An alternative and somewhat more general approach is as follows: if $f(x,y,\theta) = 0$ is the equation of a family of curves depending upon the parameter $\theta$, the envelope, if it exists, is a curve which is tangent at each point to some member of the family and each member of the family is tangent to the curve. Using the theory of partial derivatives (Spiegel, 1974) it can be shown that the equation of the envelope is given by solving simultaneously

$$f(x,y,\theta) = 0$$

and

$$\frac{\partial f}{\partial \theta}(x,y,\theta) = 0 \tag{3.6}$$

In this case

$$\frac{\partial f}{\partial \theta} = x \sec^2\theta - \frac{gx^2}{V_0^2}\sec^2\theta \tan\theta = 0 \tag{3.7}$$

$$\Rightarrow \quad \tan\theta = V_0^2/gx = a/x \tag{3.8}$$

where $a = V_0^2/g$.

Solving (3.8) and (3.1) provides us with the equation of the Enveloping Parabola:

$$2ay = a^2 - x^2$$

which is the same as (3.5). The reader may care to verify that the velocity of the projectile when it touches the envelope is perpendicular to the initial velocity.

## 3.2  WORKED EXAMPLES

1. Consider this very simple model of the shot put: a shot is thrown at a speed $V_0$ from ground level and moves solely under the action of gravity. What is its maximum range and the corresponding angle of projection?

   Solution:

   The equation of the enveloping parabola for such a shot is given by (3.5) as $2ay = a^2 - x^2$ where $a = V_0^2/g$. Since this curve divides accessible and inaccessible regions the furthest accessible point on the ground must be A (Fig. 3.3).
   Now when $y = 0$, $x = a$ since A lies on the enveloping parabola. We conclude that the maximum range $= a = V_0^2/g$. From (3.8) the corresponding angle of projection is such that $\tan\theta = 1$, i.e. $\theta = 45°$.
   There are a number of major assumptions in this model which significantly affect the range. We consider these further in Chapter 6.

2. Consider the problem of firing a cannon ball at speed $V_0$ up a plane inclined at an angle $\beta$ to the horizontal as shown in Fig. 3.4. Suppose we want to aim the cannon so that the cannon ball hits the ground at maximum range. At what angle must we fire it?

   Solution: Let OA be the maximum range required. 'A' must lie on the enveloping parabola. The coordinates of A, $(X,Y)$, must therefore satisfy

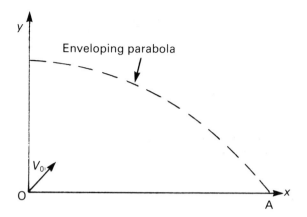

Fig. 3.3 — The shot-put problem.

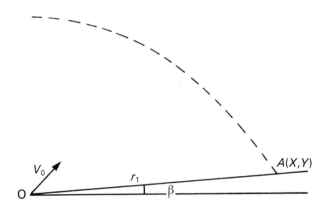

Fig. 3.4 — Cannon ball fired up an inclined plane.

$$2aY = a^2 - X^2 \tag{3.9}$$

If OA $=r_1$, then $X = r_1\cos\beta$ and $Y = r_1\sin\beta$. Thus from (3.9) we obtain

$$2ar_1\sin\beta = a^2 - r_1^2\cos^2\beta$$

Solving this equation for $r_1$ gives

$$r_1 = \frac{-2a\sin\beta \pm \sqrt{(4a^2\sin^2\beta + 4a^2\cos^2\beta)}}{2\cos^2\beta}$$

i.e.

$$r_1 = \frac{-a\sin\beta + a}{\cos^2\beta} = \frac{a(1 - \sin\beta)}{\cos^2\beta} \tag{3.10}$$

We have ignored the negative root. Why?
Now since $\tan\theta = a/X$, from (3.8) we obtain

$$\tan\theta = \frac{a\cos^2\beta}{a(1 - \sin\beta)\cos\beta} = \frac{\cos\beta}{1 - \sin\beta}$$

as the tangent of the required angle of projection.
    We can extend this problem to the case where the cannon ball is fired from a height $h$ above the ground as illustrated in Fig. 3.5.

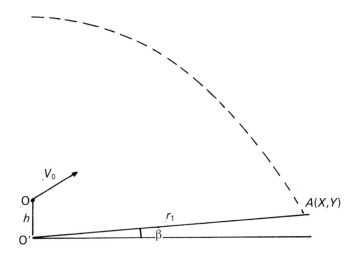

Fig. 3.5 — Cannon ball fired from an elevation $h$ up an inclined plane.

    Let O′A represent the maximum range. 'A' must lie on the enveloping parabola and so its coordinates $(X,Y)$ must satisfy

$$2aY = a^2 - X^2$$

If O′A $= r_1$ then we have $X = r_1\cos\beta$ and $Y = -h + r_1\sin\beta$. We obtain $2a(r_1\sin\beta - h) = a^2 - r_1^2\cos^2\beta$, the solution of which is

$$r_1 = \frac{-a\sin\beta \pm \sqrt{(a^2 + 2ah\cos^2\beta)}}{\cos^2\beta}$$

If we take the positive root, this is the maximum range up the inclined plane under the conditions stated. As before (3.8) gives the optimum angle of release. Note in particular that this angle depends upon the height of release — a fact which will be crucial when we attempt to standardise athletic field events in Chapter 6. If $h = 0$ the expression reduces to the formula given previously, and if $\beta = 0$ we obtain $R = \sqrt{(a^2 + 2ah)}$, which is the maximum range on a horizontal plane when the cannon ball is projected from a height $h$ above the ground. What meaning can be attached to the formula if $\beta$ is negative?

## 3.3   THE DUAL PROBLEM

In developing the theory of the enveloping paprabola we were, in fact, solving the problem of finding the maximum displacement for a given value of $V_0$. Suppose, now, we ask the question: 'What is the minimum value of $V_0$ in order that the projectile will pass through the fixed point with coordinates $(X,Y)$?'

The equation of the trajectory through $(X,Y)$ is

$$Y = X\tan\theta - \frac{gX^2\sec^2\theta}{2V_0^2} \tag{3.11}$$

Differentiate with respect to $\theta$ to obtain

$$0 = X\sec^2\theta - gX^2\sec^2\theta\left(\frac{\tan\theta}{V_0^2} - \frac{1}{V_0^3}\frac{dV_0}{d\theta}\right)$$

For a turning point, $dV_0/d\theta = 0$. Let the minimum value of $V_0$ be $V_M$. (The reader may care to convince himself that there is no maximum value.)

Then

$$X = \frac{gX^2\tan\theta}{V_M^2} \Rightarrow \tan\theta = \frac{V_M^2}{gX}$$

$$\Rightarrow \tan\theta = a/X$$

where, as usual, $a = V_M^2/g$.

Eliminating $\theta$ from (3.11) we obtain, after some elementary manipulation,

$$a^2 = X^2 + 2aY$$

This is the equation of the enveloping parabola! So $(X,Y)$ lies on the bounding parabola of the minimum value of $V_0$. We may conclude that the problem of finding a maximum displacement with a given initial speed is the same as finding the minimum speed for a given displacement.

### WORKED EXAMPLE

A boy is standing in his garden, 10 m from a tall brick wall which rises 15 m higher than his head. He intends to throw a ball over the wall. What is the minimum velocity needed to clear the wall?

Solution:

The ball must clear the point with coordinates $(10, 15)$. We have, therefore,

$$a^2 = 100 + 2a(15)$$

The solution of this equation is $a = 33.03$ which implies that $V_M = 18.2\,\text{ms}^{-1}$. Incidentally the angle of projection is $73.2°$. It is most unlikely that a boy could project a ball at $18.2\,\text{ms}^{-1}$ at such a large angle. If, however, he were to stand a greater distance from the wall, say 25 m, the minimum speed would be $21\,\text{ms}^{-1}$ at an angle of $60.5°$. This is more reasonable.

On the other side of the wall there is a conservatory with dimensions shown in Fig. 3.6. With these parameters is there a risk of breaking panes of glass in the roof?

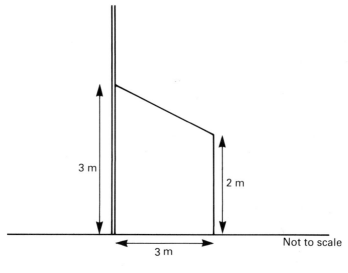

Fig. 3.6.

Assume the boy releases the ball 25 m from the wall at a height of 1.5 m.

The reader is now referred to the Percy Grainger Problem in Chapter 8 where this kind of problem is discussed in more detail.

Computer Program 1 in Appendix III illustrates an enveloping parabola.

**EXERCISES**

1. Refer to Worked Example 2 in section 3.2. Show that for maximum range, the angle of projection must bisect the angle between the plane and the vertical, when the cannon ball is fired from ground level.

2. A shot-putter releases his shot at a speed $12.6\,\text{ms}^{-1}$ from a height of $2.13\,\text{m}$. Taking $g = 10\,\text{ms}^{-2}$, write down the equation of the enveloping parabola. Use this equation to determine the maximum range of the shot. Hence find the angle of projection which achieves this range.

3. A particle is projected at an angle $\theta$ to the horizontal from the foot of a plane whose inclination to the horizontal is $\alpha$ ($\theta > \alpha$). By writing down the equation of the enveloping parabola, find the value of $\theta$ for which the range on the inclined plane is maximum. The particle is then projected from the top of the plane. Find the maximum range down the plane when the speed of projection is $U$.

4. Rework Worked Example 2 in Chapter 2 using the method described in section 3.3.

# 4

# Resisted motion

## 4.1 INTRODUCTION

So far we have only talked about a projectile moving freely under gravity. In many practical cases it is necessary to consider air resistance. We still model the projectile as a particle of mass $m$. It is conventional to write the resistive force, $\mathbf{R}$, in the form

$$\mathbf{R} = -mkV^n\hat{\mathbf{V}} \tag{4.1}$$

where $V$ is the projectile's speed, $\hat{\mathbf{V}}$ is a unit vector in the direction of motion, and $k$ and $n$ are constants. It is experimental work and the principles of fluid mechanics which will enable us to determine the values of $k$ and $n$. This is discussed in the next chapter.

## 4.2 RESISTIVE FORCE PROPORTIONAL TO THE VELOCITY

To begin with we shall consider the case where $n = 1$, i.e. the resistance is proportional to the velocity. (As we shall see in Chapter 5, this is only true when the particle is very small, moving very slowly, in a highly viscous medium (Stoke's Law). However, we shall proceed to develop the theory because we can obtain analytic solutions which can sometimes be useful approximations to the true state of affairs.)

In Fig. 4.1 a projectile has been launched from O at $t = 0$ at an angle $\theta$ to the horizontal. At some later time $t$ the projectile is at P, where $OP = r$.

The equation of motion of the particle is

$$m\ddot{\mathbf{r}} = -mkV\hat{\mathbf{V}} - mg\mathbf{j}$$

i.e.

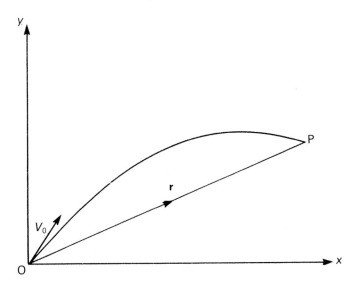

Fig. 4.1

$$\ddot{\mathbf{r}} = \begin{pmatrix} -ku \\ -g - kw \end{pmatrix} \tag{4.2}$$

where $u$ and $w$ are respectively the horizontal and vertical components of the velocity at time $t$.

Initially $\mathbf{V} = \mathbf{V}_0 = (u_0 \ w_0)^{\mathrm{T}}$ and $\mathbf{r} = \mathbf{r}_0 = (0 \ 0)^{\mathrm{T}}$.

Equation (4.2) is best solved by writing it in component form:

Horizontally: $\dfrac{du}{dt} = -ku$

$$\Rightarrow \int_{u_0}^{u} \frac{du}{u} = -k \int_0^t dt$$

so that

$$\log_e \frac{u}{u_0} = -kt$$

or

$$u = u_0 e^{-kt} \tag{4.3}$$

Hence

$$\frac{dx}{dt} = u_0 e^{-kt}$$

$$\Rightarrow \int_0^x dx = u_0 \int_0^t e^{-kt} dt$$

so that

$$x = \frac{u_0}{k}(1 - e^{-kt}) \qquad (4.4)$$

Vertically: $\dfrac{dw}{dt} = -g - kw$

$$\Rightarrow \int_{w_0}^w \frac{dw}{g + kw} = -\int_0^t dt$$

$$\Rightarrow \frac{1}{k} \log_e \left( \frac{g + kw}{g + kw_0} \right) = -t$$

so that

$$w = \frac{1}{k}(g + kw_0)e^{-kt} - \frac{g}{k} \qquad (4.5)$$

Therefore

$$\frac{dy}{dt} = \frac{1}{k}(g + kw_0)e^{-kt} - \frac{g}{k}$$

$$\Rightarrow \int_0^y dy = \int_0^t \frac{1}{k}\left( (g + kw_0)e^{-kt} - \frac{g}{k} \right) dt$$

i.e.

$$y = \frac{1}{k^2}(kw_0 + g)(1 - e^{-kt}) - \frac{gt}{k} \qquad (4.6)$$

Equations (4.4) and (4.6) are the parametric equations of the trajectory. It is possible to obtain the Cartesian equation by writing

$$t = -\frac{1}{k}\log_e\left(1 - \frac{kx}{u_0}\right)$$

from (4.4) and substituting into (4.6) to obtain

$$y = \left(w_0 + \frac{g}{k}\right)\frac{x}{u_0} + \frac{g}{k^2}\log_e\left(1 - \frac{kx}{u_0}\right)$$

If we now write $w_0 = V_0\sin\theta$ and $u_0 = V_0\cos\theta$ we obtain

$$y = x\tan\theta + \frac{gx\sec\theta}{kV_0} + \frac{g}{k^2}\log_e\left(1 - \frac{kx\sec\theta}{V_0}\right) \tag{4.7}$$

which is the equation of the trajectory. The reader should compare this with (2.9).


## 4.3  SOME STANDARD RESULTS

We will now explore some consequences of these results.

(i)  Equation (4.2) can be rewritten as

$$\frac{d\mathbf{V}}{dt} = -k\mathbf{V} - g\mathbf{j}$$

which implies the acceleration $d\mathbf{V}/dt$ is zero when $\mathbf{V} = \mathbf{V}_T = -(g/k)\mathbf{j}$. $\mathbf{V}_T$ is called the terminal velocity. It follows that any projectile, however launched, will eventually move with a constant speed vertically downwards assuming the resistance is proportional to velocity.

From (4.3) and (4.5) we obtain confirmation of this fact: if $t \to \infty$ then $u \to 0$ and $w \to -g/k$. It follows from (4.4) that $x \to u_0/k$.

So $x = u_0/k = V_0\cos\theta/k$ is an asymptote to the trajectory given by (4.7). Knowledge of $V_T$ from experimental data will provide a value for $k$.

(ii)  If in (4.7) we expand $\log_e(1 - kx\sec\theta/V_0)$ as a power series (assuming that $kx\sec\theta < V_0$, i.e. $k$ is small) we obtain

$$y = x\tan\theta + \frac{gx\sec\theta}{kV_0} + \frac{g}{k^2}\left(\frac{-kx\sec\theta}{V_0} - \frac{k^2x^2\sec^2\theta}{2V_0^2} - \frac{k^3x^3\sec^3\theta}{3V_0^3}\cdots\right)$$

i.e.

$$y = x\tan\theta - \frac{gx^2\sec^2\theta}{2V_0^2} - \frac{gkx^3\sec^3\theta}{3V_0^3}(+\text{ terms involving higher powers of } k)$$

Now as $k \to 0$,

$$y \to x\tan\theta - \frac{gx^2\sec^2\theta}{2V_0^2}$$

which is the equation of the trajectory in the absence of air resistance. This shows that these results are consistent with those of Chapter 2.

(iii) We now present some results about the range and maximum height, as we did in Chapter 2.
   We have

$$x = \frac{u_0}{k}(1 - e^{-kt})$$

$$y = \frac{1}{k^2}(kw_0 + g)(1 - e^{-kt}) - \frac{gt}{k}$$

and

$$w = \frac{1}{k}(g + kw_0)e^{-kt} - \frac{g}{k}$$

In order to find the range on a horizontal plane through the point of projection we first find the value of the time $\tau_1$, when $y = 0$, and substitute the result to find the corresponding value of $x$ which we will denote by $R$.
   When $y = 0$ we have

$$\frac{(kw_0 + g)(1 - e^{-k\tau_1})}{k^2} = \frac{g\tau_1}{k}$$

so that

$$\tau_1 - \frac{(kw_0 + g)(1 - e^{-k\tau_1})}{kg} = 0$$

This equation cannot be solved analytically, and a numerical method must be used.

To this end, and as an example, we put $k = 0.1$†, $g = 10\,\mathrm{ms}^{-2}$, and $u_0 = w_0 = 7.07\,ms^{-1}$, *i.e.* $\theta = 45°$.

The equation becomes

$$f(\tau_1) = \tau_1 - 10.707(1 - e^{-0.1\tau_1}) = 0$$

This can be solved by a simple iteration method or the Newton–Raphson method. In either case a starting value of 1.4 should be used. The value of $\tau_1$ is $1.382\,\mathrm{s}$. This gives a value for the range, $R$, of $9.13\,\mathrm{m}$. This should be compared with the value of $10\,\mathrm{m}$ in the absence of air resistance. Note, however, that although 45° is the optimum angle of projection in the absence of air resistance, this is not so in the resisted case, so $9.13\,\mathrm{m}$ is not the maximum range for the given initial speed.

It is possible to find an analytic expression for the greatest height $H$. We put $w = 0$ to find the time $\tau_2$ for the projectile to reach its maximum height and then substitute $\tau_2$ into the formula for $y$.

$$w = 0 \Rightarrow (g + kw_0)e^{-k\tau_2} = g$$

$$\Rightarrow e^{-k\tau_2} = \frac{g}{g + kw_0}$$

so that

$$\tau_2 = \frac{1}{k}\log_e \frac{(g + kw_0)}{g}$$

Substitution into the formula for $y$ gives an expression for $H$:

$$H = \frac{1}{k^2}\left(kw_0 - g\log_e \frac{(g + kw_0)}{g}\right)$$

Using the same values as above we obtain $\tau_2 = 0.683\,\mathrm{s}$. We note that this is less than $\tau_1/2$ so that the downward journey takes longer.

The value of $H$ is just under $2.4\,\mathrm{m}$. In the absence of air resistance the greatest height would be $2.5\,\mathrm{m}$.

(iv) We stated above that $x = V_0\cos\theta/k$ is an asymptote to the trajectory. A typical one is sketched in Fig. 4.2. Clearly the trajectory is not a parabola.

Computer Program 2 in Appendix III will plot on the screen trajectories for the case $V_0 = 10\,\mathrm{ms}^{-1}$ and $k = 0.1\,\mathrm{s}^{-1}$ for a range of values of $\theta$. It will be seen that the trajectories have an envelope (cf. the Bounding Parabola). Although it is possible to find its Cartesian equation, we choose to adopt a numerical approach as follows:

† The justification for this value of $k$ is given for one particular sphere in Chapter 7.

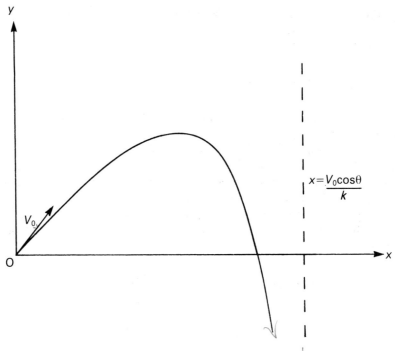

Fig. 4.2 — A typical trajectory when $R \propto V$.

as stated in Chapter 3, the envelope of a family of curves, if it exists, is found by eliminating $\theta$ between the equations

$$f(x,y,\theta) = 0 \text{ and } \frac{\partial f}{\partial \theta}(x,y,\theta) = 0$$

From (4.7) we can write $f(x,y,\theta)$ as

$$f(x,y,\theta) = x\tan\theta + \frac{gx\sec\theta}{kV_0} + \frac{g}{k^2}\log_e\left(1 - \frac{kx\sec\theta}{V_0}\right) - y = 0$$

and so

$$\frac{\partial f}{\partial \theta} = x\sec^2\theta + \frac{gx\tan\theta\sec\theta}{kV_0} - \frac{gx\tan\theta\sec\theta}{k(V_0 - kx\sec\theta)}$$

Therefore

$$x\sec\theta + \frac{gx\tan\theta}{kV_0} - \frac{gx\tan\theta}{k(V_0 - kx\sec\theta)} = 0$$

This equation can be solved for $x$ to give

$$x = \frac{V_0^2}{V_0 k\sec\theta + g\tan\theta} \qquad (4.8)$$

Here we note that if $k = 0$ then $x = V_0^2/g\tan\theta$ as before. Putting $V_0 = 10\,\mathrm{ms}^{-1}$, $k = 0.1\,\mathrm{s}^{-1}$, and $g = 10\,\mathrm{ms}^{-2}$ as above we obtain

$$x = \frac{100}{\sec\theta + 10\tan\theta} \qquad (4.9)$$

The envelope can be constructed by solving (4.9) for $x$ for a range of values of $\theta$, and then substituting $x$ and $\theta$ into (4.7) to find $y$. Values of $\theta, x$ and $y$, obtained by this method, are given in Table 4.1.

**Table 4.1 — Data for the determination of the envelope**

| $\theta°$ | $x$ | $y$ |
|---|---|---|
| 10.000 | 35.988 | − 83.034 |
| 20.000 | 21.259 | − 22.514 |
| 30.000 | 14.434 | − 7.322 |
| 40.000 | 10.313 | − 1.314 |
| 50.000 | 7.422 | 1.617 |
| 60.000 | 5.176 | 3.206 |
| 70.000 | 3.290 | 4.093 |
| 80.000 | 1.601 | 4.549 |

Any trajectory given by (4.7) may be labelled $T_\theta$. The values of $x$ and $y$ above are points on the boundary between accessible (A) and inaccessible (I) regions. Write $(x,y)$ as $P_\theta$ and the angle between $\overline{OP_\theta} = \mathbf{r}_\theta$ and the horizontal as $\phi_\theta$. Then for any $\phi_\theta$, $\mathbf{r}_\theta$ is the maximum displacement vector, as illustrated in Fig. 4.3.

A graph of the boundary is given in Fig. 4.4 for $V_0 = 10\,\mathrm{ms}^{-1}$ and $k = 0.1$. This envelope could be useful in maximum range problems where $R \propto V$ is a reasonable approximation. (See Baiera's problem in Chapter 7.)

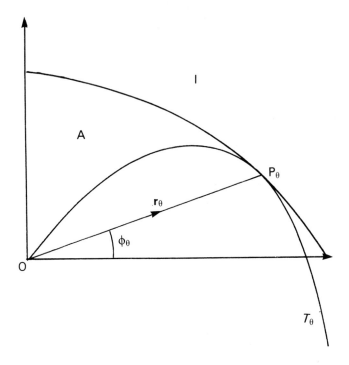

Fig. 4.3

Murphy (1979) has shown that in the case $R \propto V$, the equation of the enveloping curve is

$$y^* = \frac{\sqrt{\{1+x^{*2}(1-\varepsilon^2)\}}}{\varepsilon} + \frac{1}{\varepsilon^2}\log_e\left[\frac{1-\varepsilon\sqrt{\{1+x^{*2}(1-\varepsilon^2)\}}}{1-\varepsilon^2}\right]$$

where

$$y^* = \frac{y}{V_0^2/g}, \quad x^* = \frac{x}{V_0^2/g}, \quad \text{and} \quad \varepsilon = \frac{kV_0}{g}$$

Incidentally it is still true that the velocity of the projectile when it touches the enveloping curve is perpendicular to its initial velocity.

## 4.4   RESISTIVE FORCE PROPORTIONAL TO THE SQUARE OF THE VELOCITY

We now consider the case of the resistive force being proportional to the square of the speed. As will be explained in detail in the next chapter, this law is applicable to many practical examples of projectile motion.

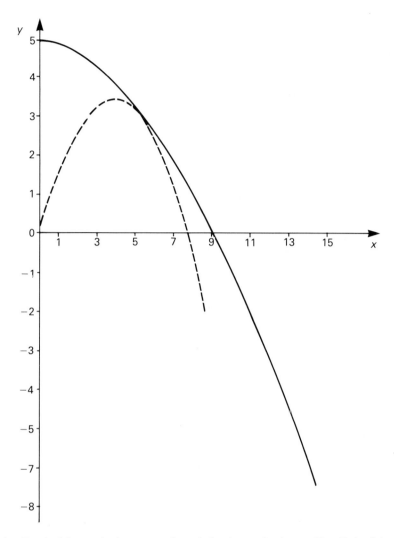

Fig. 4.4 — Graph of the enveloping curve and a typical trajectory for the case $V_0 = 10$, $k = 0.1$.

We can write (4.1) as

$$\mathbf{R} = -mkV^2\hat{\mathbf{V}}$$

$$= -mk|\mathbf{V}|^2\frac{\mathbf{V}}{|\mathbf{V}|} = -mk|\mathbf{V}|\mathbf{V}$$

In Fig. 4.1 a projectile has just been launched from O at $t = 0$ at angle $\theta$ to the horizontal. At some later time $t$ the projectile is at P where OP $= r$. As before, $u$ and $w$ are the horizontal and vertical components of the velocity $\mathbf{V}$.

$$V = |\mathbf{V}| = (u^2 + w^2)^{\frac{1}{2}}$$

The equation of motion of the projectile is

$$\ddot{\mathbf{r}} = \begin{pmatrix} -k(u^2 + w^2)^{\frac{1}{2}}u \\ -g - k(u^2 + w^2)^{\frac{1}{2}}w \end{pmatrix}$$

i.e.

$$\frac{d^2x}{dt^2} = \frac{du}{dt} = -k(u^2 + w^2)^{\frac{1}{2}}u$$

$$\frac{d^2y}{dt^2} = \frac{dw}{dt} = -g - k(u^2 + w^2)^{\frac{1}{2}}w \qquad (4.10)$$

These equations are not independent and there is no analytical way of solving them. However, there are several numerical techniques available. One of these (the Runge–Kutta method) is used in Computer Program 3 in Appendix III which illustrates the trajectory of such a projectile. If the projectile were moving vertically then the last equation would reduce to

$$\frac{d^2y}{dt^2} = \frac{dw}{dt} = -g - k|w|w$$

The acceleration $dw/dt$ is zero when the weight is balanced by the resistance, i.e.

$$-g = k|w_T|w_T$$

where $w_T$ is the terminal velocity. Thus $w_T$ is given by $-\sqrt{g/k}$. If $k = 0.4$ then this terminal velocity is $5\,\mathrm{ms}^{-1}$ (about 11 mph), a fact much appreciated by parachutists.

## 4.5   INTRINSIC COORDINATES

An alternative approach is to use intrinsic rather than Cartesian coordinates. Intrinsic coordinates are moving coordinates which are located at the position of the projectile (P), along the tangent (i.e. the instantaneous direction of motion) and along the normal at that point. At any time the tangent makes an angle $\alpha$ with the horizontal. The forces acting on the projectile are

$$-mkV^2 - mg\sin\alpha \qquad\qquad\qquad \text{along the tangent}$$

and

$$- mg\cos \alpha \qquad\qquad\qquad\qquad \text{along the normal}$$

This is illustrated in Fig. 4.5

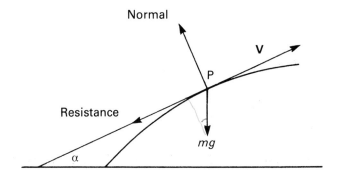

Fig. 4.5 — Intrinsic coordinates.

The equations of motion are

$$\frac{dV}{dt} = - kV^2 - g\sin\alpha \qquad\qquad (4.11)$$

$$V\frac{d\alpha}{dt} = - g\cos\alpha$$

(See, for example, Quadling & Ramsey (1966).)
Now

$$\frac{dV}{dt} = \frac{dV}{d\alpha}\frac{d\alpha}{dt}$$

Therefore

$$\frac{dV}{d\alpha}\left(\frac{- g\cos\alpha}{V}\right) = - kV^2 - g\sin\alpha$$

$$\Rightarrow \frac{dV}{d\alpha}\cos\alpha - V\sin\alpha = \frac{kV^3}{g} \qquad\qquad (4.12)$$

If we now make the substitution $u = V\cos\alpha$, then (4.12) becomes

$$\frac{du}{d\alpha} = \frac{ku^3}{g\cos^3\alpha}$$

so that

$$\int \frac{du}{u^3} = \int \frac{kd\alpha}{g\cos^3\alpha}$$

$$\Rightarrow \frac{-1}{2u^2} = \int \frac{kd\alpha}{g\cos^3\alpha}$$

Writing $\sec^3\alpha = \sec\alpha \sec^2\alpha$, this integral can be evaluated by parts and some lengthy manipulation to give

$$\frac{k}{2g}(\sec\alpha \tan\alpha + \log_e(\sec\alpha + \tan\alpha)) + \text{constant}$$

When the initial conditions $\alpha = \theta$ when $V = V_0$ are substituted, we obtain

$$V = \left\{ \frac{\cos^2\alpha}{V_0^2\cos^2\theta} - \frac{k\cos^2\alpha}{g}\left[ \log_e\left(\frac{\sec\alpha + \tan\alpha}{\sec\theta + \tan\theta}\right) + \frac{\sin\alpha}{\cos^2\alpha} - \frac{\sin\theta}{\cos^2\theta} \right] \right\}^{-\frac{1}{2}}$$

We now have an expression for the projectile's velocity in terms of its direction of motion.

From (4.11) we have

$$\frac{dt}{d\alpha} = \frac{-V\sec\alpha}{g}$$

Also

$$\frac{dx}{d\alpha} = \frac{dx}{dt}\frac{dt}{d\alpha} = V\cos\alpha \frac{dt}{d\alpha} = -\frac{V^2}{g} \tag{4.13}$$

Similarly

$$\frac{dy}{d\alpha} = \frac{dy}{dt}\frac{dt}{d\alpha} = V\sin\alpha \frac{dt}{d\alpha} = -\frac{V^2\tan\alpha}{g}$$

From (4.13) we obtain

$$x = \frac{-1}{g} \int_\theta^\alpha V^2 \, d\alpha$$

$$y = \frac{-1}{g} \int_\theta^\alpha V^2 \tan\alpha \, d\alpha$$

$$t = \frac{-1}{g} \int_\theta^\alpha V \sec\alpha \, d\alpha$$

These equations have been solved by Tan (1987) in connection with the modelling of shuttlecock trajectories in badminton. We reproduce his results in Fig. 4.6. They can also be produced by modifying Computer Program 3 in Appendix III (take $k = 0.212 \, \text{m}^{-1}$ and $m = 5.12 \times 10^{-3} \, \text{kg}$).

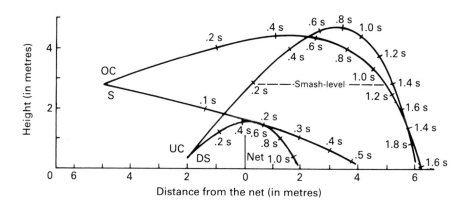

Fig. 4.6 — Shuttlecock trajectories in badminton. OC = overhand clear launched at 40 m/s, 20°; S = smash, 60 m/s, − 12°; UC = underhand clear, 30 m/s, 50°; DS = drop shot, 9 m/s, 45°. (From 'Shuttlecock trajectories in badminton', by A. Tan, in *Mathematical Spectrum*, with permission.)

## 4.6  WORKED EXAMPLE

This example will illustrate some techniques and enable the reader to tackle the following exercises.

A projectile is fired at 36° to the horizontal with a speed of 20 ms$^{-1}$, in a medium in which the resistance varies as the velocity. It is observed that it is moving horizontally one second after firing. How far will it travel horizontally if the point of firing is 1 m above the horizontal ground?

Solution: The basic results we need to solve this problem are

$$x = \frac{u_0}{k}(1 - e^{-kt}) \tag{4.14}$$

$$y = \frac{1}{k^2}(kw_0 + g)(1 - e^{-kt}) - \frac{gt}{k} \tag{4.15}$$

and

$$w = \frac{1}{k}(g + kw_0)e^{-kt} - \frac{g}{k} \tag{4.16}$$

We first need to obtain a value for $k$. When the projectile is moving horizontally then $w = 0$ and we know that $t = 1$. So (4.16) becomes

$$(10 + 11.76k)e^{-k} - 10 = 0$$

i.e.

$$k = \log_e\left(\frac{10 + 11.76k}{10}\right)$$

Starting with $k_0 = 0.3$, a simple iterative method will give $k = 0.316$. We now require the time taken for the projectile to reach ground level, i.e. $y = -1$. It is easily verified from (4.15) that $t$ is obtained from the equation

$$t = 4.34(1 - e^{-0.316t}) + 0.032$$

Starting with $t_0 = 2.2$, simple iteration leads to $t = 2.221$. Hence from (4.14) the horizontal displacement is 25.82 m.

## EXERCISES

1. The following is a report from the *Daily Telegraph*, 100 years from now.

### WORCESTER, APRIL 1st 2088

Landing, after a return journey of over 10 years, at the new Interplanetary Station alongside the River Severn, Professor Botham, whose only claim to fame up to now was that he is the grandson of the legendary cricketer of the last century, Ian Botham, has finally arrived back on Earth from his historic visit to $\alpha$-centauri3. Speaking to our Projectile Correspondent he said that the radius of the planet was 2.5 times smaller than that of the Earth but 'made of the same sort of stuff'. He went on: 'My grandson, who played on this very spot, would not have enjoyed cricket out there. You could breathe the atmosphere O.K. but it was so dense!'

Stoke's Law, old as it is, was certainly obeyed. 'I dropped a stone from shoulder height, say a metre , and it took 0.8 s to reach the ground! We had a young athlete on board who could throw a cricket ball 100 m on Earth where the resistance is negligible. He had a go on $\alpha$-centauri3. You'll never guess how far it went'.

How far do you calculate it went?

2. Survivors from a shipwreck are on a very small island. A light aircraft is commissioned to drop essential supplies to them by parachute. The idea is that the plane will fly horizontally at a steady speed in a direction which will take it directly over the middle of the island. The pilot has been told that the parachute will open immediately and because of its canopy the parcel will suffer a resistance proportional to the square of the speed in the vertical direction. Horizontally the resistance is negligible. Where should the pilot release the package? Investigate. (Assume reasonable values for the height and speed of the aeroplane and take $k = 0.4$.)

3. Naturally the Nantwich Sub-Aqua Athletics Association play to different rules from their land-based colleagues. Because they compete in 12 feet of water, the shot-putters have devised two sets of rules. In the first it is a foul if the shot breaks the surface of the water and in the second it is a foul if it does not. The athletes are as tall and as strong as the 'dry' ones. What do you consider might be considered to be a good throw under each set of rules?. Incidentally, the association's mathematicians agree that, whereas in the air the shot undergoes no resistance, in water, for simplicity, the drag is proportional to the speed (take $k = 0.3$).

# 5

# Aerodynamic aspects of projectile motion

## 5.1 THE AERODYNAMIC FORCE

Whenever a projectile moves through the air it experiences a resistance called the aerodynamic force. (This is in addition to the weight and the buoyancy force or upthrust which is equal to the weight of the air displaced by the body. Generally the upthrust is negligible.) To facilitate understanding it is conventional to resolve the aerodynamic force into two components, one of which is perpendicular to the direction of motion of the centre of gravity of the projectile, and is known as *lift*, and the other which is in the opposite direction to the direction of motion, and is known as *drag* (Fig. 5.1).

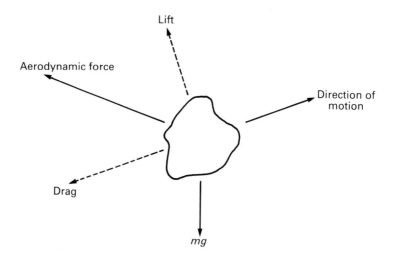

Fig. 5.1 — The aerodynamic force.

If the projectile has an irregular shape then, in general, the drag will be much larger than the lift and the major effect of the aerodynamic force is to retard the motion. However, by carefully designing the shape of the body it is possible to reduce the drag and simultaneously increase the lift. In this case, the effect of the aerodynamic force is to help sustain the motion. Knowledge of the forces on such 'aerodynamic bodies' is of great importance in the design of aircraft, propellers, javelins, etc.

On occasions it will be useful to consider the projectile as being fixed and the air as being in motion rather than the other way around. The reader should note that the flow past a fixed body in an air stream which is steady and uniform at infinity is exactly equivalent to the flow produced by a body moving steadily through the air which is at rest at infinity. Nothing essential is lost by changing the frame of reference.

## 5.2   THE ORIGINS OF THE AERODYNAMIC FORCE

For our purposes the total aerodynamic force on a body moving through the air can be considered to arise from two sources. Consider a small portion of the surface of the body as illustrated in Fig. 5.2.

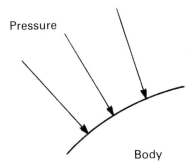

Fig. 5.2 — The pressure force.

The air close to the body exerts a normal pressure force on the surface at every point. The total pressure force on the body can be found by integrating the pressure force over the whole body surface. In addition, friction provides a tangential force at each point which opposes the motion of the air past the projectile (Fig. 5.3).

The total frictional force is found by integrating the tangential force over the whole surface. It is the combination of the pressure and frictional forces which gives rise to the total aerodynamic force. As we have stated, the total aerodynamic force is made up of contributions from all over the body surface. It is nevertheless convenient mathematically to regard this force as acting at a single point, the position of which is chosen so that the effect is the same. This point is known as the *centre of pressure* and in general it will not coincide with the centre of gravity.

## 5.3   THE REYNOLDS NUMBER, Re

What are apparently different types of flow around geometrically similar objects are said to be dynamically similar if they have roughly the same Reynolds number, Re.

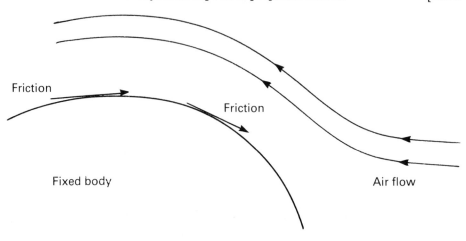

Fig. 5.3 — The frictional force.

The derivation and background of this important quantity is beyond the scope of this book and the interested reader should refer to a fluid mechanics text for more details. The Reynolds number of a flow about an object is defined to be

$$\mathrm{Re} = \frac{\rho V L}{\mu}$$

where $\rho$ is the density of the fluid, $V$ is the speed of the flow, $L$ is a typical length scale and $\mu$ is the viscosity of the fluid.

For example, the Reynolds number for a small plastic sphere of diameter $0.6\,\mathrm{cm}$ moving at $0.55\,\mathrm{ms}^{-1}$ in water of viscosity $0.89 \times 10^{-3}\,\mathrm{kgm}^{-1}\mathrm{s}^{-1}$ and density $1000\,\mathrm{kgm}^{-3}$ is 3708. A geometrically similar helium-filled balloon of diameter $1\,\mathrm{m}$ moving at $0.056\,\mathrm{ms}^{-1}$ in air of viscosity $1.81 \times 10^{-5}\,\mathrm{kgm}^{-1}\mathrm{s}^{-1}$ and density $1.205$ $\mathrm{kgm}^{-3}$ has Reynolds number 3728. Therefore the two flows are dynamically similar. This concept is most useful since flows which are similar behave in the same way so that the flow around the balloon under the conditions given will be identical to the flow about the plastic ball. This principle is widely used as means of obtaining information about an unknown flow from dynamically similar models. The reason why Re is so important here is that a flow exhibits different types of behaviour at different Reynolds numbers. Loosely speaking, a flow with a low Re has a highly viscous behaviour whereas a high Re implies inviscid behaviour. In particular, the law of drag changes with Reynolds number. Take for example the flow past a smooth sphere: at very low Reynolds number (Re $< 0.2$) the drag force is found experimentally to be proportional to the speed (Stoke's law). At high Re ($500 < \mathrm{Re} < 10^5$) this force is found to be proportional to the square of the speed, although this is only approximately true. At intermediate values, more complex behaviour ensues.

## 5.4   THE COEFFICIENT OF DRAG, C$_D$

Both the pressure drag and the frictional drag can be theoretically calculated provided knowledge of the pressure distribution around the body and the frictional stress distribution on the surface is available. The determination of these quantities can be difficult, especially for objects having complicated shapes, and so the total drag is generally measured experimentally. This is achieved by measuring the forces on an object when it is placed in a wind tunnel.

It is conventional to express the drag in the form

$$D = \tfrac{1}{2}C_D\rho V^2 A$$

where $A$ is the area of the projection of the body on a plane normal to the air stream at infinity. $C_D$ is known as the *coefficient of drag*. For a smooth sphere moving with speed $V$ in still air we know that at high Re the drag is proportional to the square of the speed, so that $C_D$ is independent of $V$. At low Re the drag is proportional to $V$, so that $C_D$ must be inversely proportional to $V$. At very high Re the drag coefficient shows a further reduction as illustrated in Fig. 5.4.

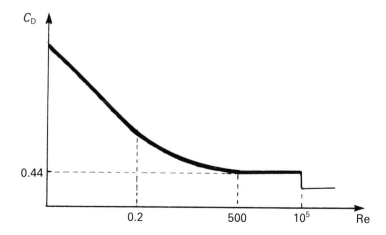

Fig. 5.4 — Variation of $C_D$ with Re.

The rate of decrease of drag coefficient in this range can be so great as to cause an overall reduction in drag even as $V$ increases. The explanation of this phenomenon lies in the behaviour of the boundary layer. This is a thin layer of fluid close to the body surface in which viscous effects are very important. In the intermediate range of Re, this boundary layer separates from the surface at about 80° from the front of the sphere. This produces a wide, separated wake in which the pressure is low (Fig. 5.5(a)).

This in turn causes a large drag. When the Reynolds number exceeds a certain critical value the laminar, or smooth, flow in the boundary layer becomes turbulent, the effect of which is to delay the separation (Fig. 5.5(b)). This causes a narrower

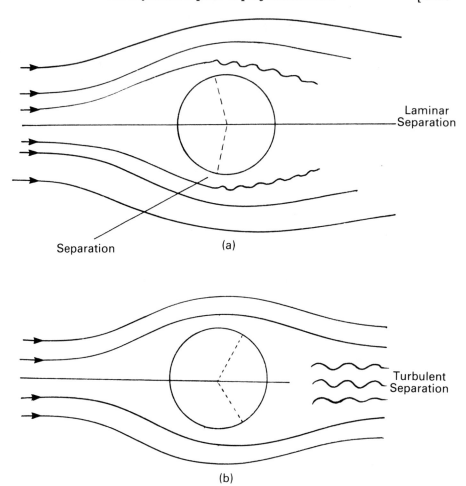

Fig. 5.5(a),(b) — Laminar and turbulent wakes.

low-pressure wake and hence reduced drag. If the free-stream flow is not particularly smooth or the surface of the body is rough, then the critical Re at which turbulence occurs is reduced. This means that at a relatively high Re when one might expect a rough sphere to experience more drag than a smooth one, the reverse is in fact the case if the flow around the rough sphere becomes turbulent.

It is apparent from Fig. 5.6 that there is a range of Re for which a smooth sphere will experience more drag than a rough one. In some applications the boundary layer is deliberately made turbulent. This is the purpose of the dimples on a golf ball — the boundary layer is made turbulent and the drag is reduced. This tends to increase the range.

At very high speeds (i.e. when the speed of the projectile approaches the speed of sound waves) the coefficient of drag is no longer independent of $V$ but rises suddenly before decreasing in the supersonic region.

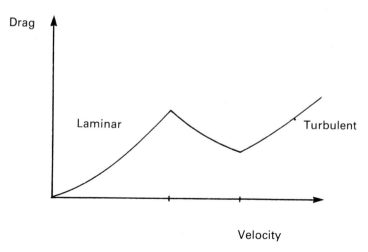

Fig. 5.6 — Variation of drag force with velocity.

The drag force can be changed by altering the attitude of the body as well as its speed, since this may affect the projected area $A$. Consider the aerodynamically shaped object in Fig. 5.7.

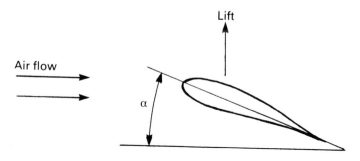

Fig. 5.7 — Angle of attack.

The angle $\alpha$ between the direction of flow and a line fixed in the body is known as the angle of attack. This parameter will be useful in Chapter 10.

## 5.5  GENERATION OF LIFT

As in section 5.4 it is conventional to write the lift in the form $L = \frac{1}{2}C_L\rho V^2 A$ where $C_L$ is called the *coefficient of lift*. Generally its value must be determined experimentally. Careful design of a body enables large amounts of lift to be generated and in this way the flight of an object is sustained. For flow past aerodynamically shaped bodies it is possible to show that the lift is related to the angle of attack $\alpha$. For simple shapes it is possible to calculate this relationship. For example, lift is proportional to $\sin \alpha$ for certain types of aerofoil provided $\alpha$ remains below a critical value. To increase lift, the angle of attack should therefore be increased. If $\alpha$ exceeds the critical value, the lift can fall rapidly and a phenomenon known as stall occurs. This behaviour is very complex and can be catastrophic. It is worth noting that negative lift will occur when the angle of attack is negative (Fig. 5.8).

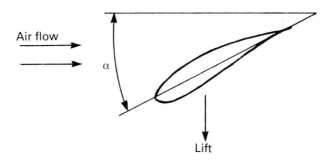

Fig. 5.8 — Negative lift.

Even if a body is not aerodynamically designed, lift can still be generated in other ways. If a cylinder is rotated in a flow with the axis of rotation perpendicular to the flow, fluid on one side can be dragged around. This causes a pressure difference on opposite sides of the cylinder and so a resultant transverse force can occur. (Incidentally this effect has been used in the Fletner Rotor ship.) When a sphere moves through the air, the flow separates from the surface as shown in Fig. 5.5. If the sphere is not spinning as it moves, there is no lift generated, but if the sphere is spinning, the points of separation of the boundary layer and consequently the position of the wake move also. This too gives rise to pressure variations and hence transverse forces, which can often be put to good use by accomplished sportsmen. Variations in roughness of the spinning surface can additionally affect the separation positions and consequently the lift and drag. Cricketers have known for some time that the behaviour of a new ball, which is smooth (apart from the seam), differs from an older, roughened one.

## 5.6   WIND

In the foregoing work we have assumed that no wind is present. Suppose we now want to modify the equations to take account of the motion of the air around a spherical projectile. Let the air have velocity $\mathbf{V}_A = (u_A \, w_A)^T$. The only way this modification affects the motion is through the drag term, since the drag force is opposite in direction to the velocity of the projectile relative to the moving air. Thus if **V** is the velocity of the projectile relative to a stationary coordinate system, the drag is in the direction shown in Fig. 5.9.

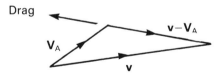

Fig. 5.9.

The drag force becomes

$$\tfrac{1}{2}\rho C_D A \,|\mathbf{V}_A - \mathbf{V}|\,(\mathbf{V}_A - \mathbf{V})$$

$$= \tfrac{1}{2}\rho C_D A \,|\mathbf{V}_A - \mathbf{V}|\begin{pmatrix} u_A - u \\ w_A - w \end{pmatrix}$$

What we have done in this and the preceding chapters is to provide a theoretical framework for the solution of projectile problems. Some of these will not involve aerodynamic forces; others will. In the latter case we will need to consider such forces on two types of object. The first is a sphere: when the sphere is not spinning there is no mechanism for generating lift. On the other hand, when it is spinning, lift is present. The second object is a javelin, and lift is generated owing to its aerodynamic shape.

## EXERCISE

The Nantwich Sub-Aqua Athletics Team are taking part in a shot-put competition in a local swimming baths. The shot is a smooth iron sphere of mass 7.257 kg and radius 6 cm. The viscosity of water is $1.311 \times 10^{-3}$ Pa s and its density is 1000 kgm$^{-3}$. One particular shot is launched at an angle of 30° at a speed of 20 ms$^{-1}$. Assuming that the shot remains under water calculate how far it travels assuming it is released from a height of 2 m above the bottom of the pool. (Assume that $C_D$ = constant = 0.4 and the resistance is proportional to the square of the speed.)

Hint: the upthrust is now significant. Modify Computer Program 8 in Appendix III to include upthrust, which is a constant vertical force.

# 6

# Standardising the shot

## 6.1 INTRODUCTION

The origins of this Olympic field even go back to the ancient Celtic game of hurling the stone. In the nineteenth century, British military athletics groups adapted the event using a cannon ball rather than a stone, and this was soon taken up by civilian athletes. The International Amateur Athletics Federation recognised the first world record in 1876. It was achieved by the American athlete J. M. Mann, who putted the shot almost 31 feet (9.4 m!). Improvements in technique led to vast increases in recorded ranges. In 1909 another American broke the 15 m barrier, and as time went on, athletes were fascinated by what became known as the '60 foot' barrier. The concern was analogous to the so-called 4 minute mile barrier. Interestingly it was only 2 days after Dr Roger Bannister's historic 4 minute mile at Oxford in May 1954 that Parry O'Brien broke the '60 foot' barrier for the shot. The current world record is 23 m.

In the modern event the athlete must project the shot from the interior of a circle of radius 2.1 m. The range is measured from the centre of the circle to the first point at which the shot touches the ground, provided that the point lies within a marked 65° sector of a circle centred at the throwing circle. In putting the shot, the athlete holds the shot against his shoulder. He springs forward off his rear leg, simultaneously leaping and straightening his legs, trunk and arms, and thus thrusts the shot forward (Fig. 6.1).

For National and International events the shot used is a sphere made of iron (or any other metal not softer than brass) of mass 7.2 kg. Smaller shots are used for women's or junior events.

Many of the recent attempts (Burghes *et al.*, 1982; Lichtenberg & Wills 1978) to model the flight of a shot have, for obvious reasons, concentrated upon the maximum range. In this chapter we will show that a more efficient method for solving problems of this type is provided by the 'enveloping parabola', introduced in Chapter 3. The

Fig. 6.1 — The shot put. Reproduced with permission, from *Applying Mathematics,* p. 98, by
Burghes *et al.* (Ellis Horwood, Chichester, 1982).

use of the enveloping parabola is desirable in any projectile problem involving maximum displacement from the origin.


## 6.2   SHOT-PUT ANALYSIS

Following Burghes *et al.* (1982) we treat the shot as a point mass moving without air resistance and we do not try to study all aspects of the shot-putter's style. We assume that in a competition, at whatever level, the shot-putter will have training and practice. Furthermore he will have practised on the field where the competition will be held. The athlete will project the shot in such a way as to maximise his put, i.e. at the greatest speed, $V_0$, and at the 'best' angle of projection for his height. Recall from Chapter 3 that the optimum angle depends upon the height of release. Our model using the enveloping parabola will therefore not be suitable for athletes who have the wrong technique or who have had no practice on the field before. (The point of this last remark will be made clear in the next section.) Let the shot be projected from the origin at a height $h$ above the ground at speed $V_0$ (Fig. 6.2).

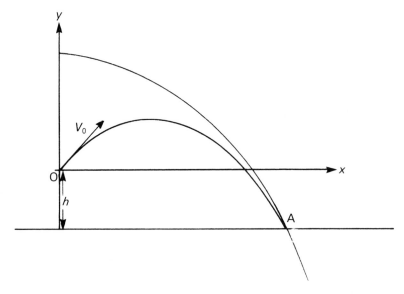

Fig. 6.2 — The shot-put problem.


If the shot hits the ground at A then its range is $R$. If the shot-putter has 'done his best' then A lies on the enveloping parabola and the coordinates of A $(R, -h)$ will satisfy (3.5), i.e.

$$2ay = a^2 - x^2$$

where $a = V_0^2/g$

$$\Rightarrow -2ah = a^2 - R^2$$

therefore

$$R = \sqrt{(a(a + 2h))} \tag{6.1}$$

Realistic values of the parameters $h$ and $a$ are $h = 2$ m, $a = 22.94$ m for adult shot-putters. Substituting in (6.1) we obtain $R = 24.86$ m and from (3.8) $\theta = 42.7°$. (This value of $R$ is somewhat larger than the current world record owing principally to the neglect of air resistance in our model.) It must be pointed out that on the basis of this model there is no single maximum range or single optimum angle of projection. Both of these depend upon $h$. The effect is small but nevertheless important in world record field events. If there were two equally skilled and strong shot-putters, one of whom released his shot at a height of 2 m and the other at 1.5 m, then the ratio of their maximum ranges would be 1.02:1 when $V_0 = 15$ ms$^{-1}$. This corresponds to a difference of almost half a metre, which clearly makes the difference between gaining an Olympic or world record, or not. Clearly the height of a tall athlete endows him with an advantage over a smaller competitor. Games' organisers may therefore wish to eliminate such an unfair variable and we consider one way of doing this in the next section.

## 6.3   A SUGGESTION FOR STANDARDISATION

Our model has shown that the maximum range for a projectile over a horizontal plane is given by (6.1) under the assumption that the thrower achieves the optimal angle of projection. The method of the enveloping parabola would be useful if athletes wished to standardise world records for unfair variables. Since $R = \sqrt{a(a + 2h)}$ and assuming that a shot-putter always produces his best speed of projection, then the only two variables are $h$ and $g$. Because the Earth is not spherical, $g$ varies with latitude, and to a first approximation is given by

$$g \doteq 9.8062 - 0.02587 \cos 2\lambda$$

where $\lambda$ is geographical latitude (Allen, 1963). It would appear that a tall athlete at the equator has an advantage over an equally skilled but short athlete at a higher latitude. We develop a method for standardising all throws to a value of $g = 10$ ms$^{-2}$ and $h = 0$, for ease of calculation.

### (i)   Standardisation for height
Suppose an athlete throws a distance $R_1$ from a height $h_1$. Then, referring to Fig. 6.2, the coordinates of A are $(R_1, -h_1)$ and A lies on the enveloping parabola, $2ay = a^2 - x^2$, where $a = V_0^2/g$.

$$\Rightarrow a^2 = R_1^2 - 2ah_1$$

so that $a^2 + 2ah_1 - R_1^2 = 0$

$$\Rightarrow a = -h_1 + \sqrt{(h_1^2 + R_1^2)} \qquad (6.2)$$

Let OA′ be the maximum range when $h = 0$. Now if $h = 0$ then, under the assumption that the shot is put in such a way as to maximise its range, A′$(R,0)$ is on the new enveloping parabola. Hence

$$R = a$$

$$\Rightarrow R = \sqrt{h_1^2 + R_1^2} - h_1 \qquad (6.3)$$

As an example, suppose an athlete released his shot at $h = 2$ m and achieved a maximum range of 21 m. The standardised range would be 19.10 m. This means that the athlete throwing from a height 2 m with his optimum trajectory would, if reduced to zero height, have a range of 19.10 m but with a different trajectory. This standardisation procedure ensures that short shot-putters are not disadvanataged, and the only important variables remaining would be the speed of projection and the skill of the athlete to produce this speed.

### (ii)   Standardisation for gravity
As before,

$$a = -h_1 + \sqrt{h_1^2 + R_1^2}$$

so

$$\frac{V_0^2}{g} = -h_1 + \sqrt{h_1^2 + R_1^2}$$

So our new value for $a$ is

$$\frac{V_0^2}{10} = \frac{ag}{10} \qquad (6.4)$$

The new range $R$

$$= \sqrt{\frac{ag}{10}\left(\frac{ag}{10} + 2h\right)}$$

standardised to $g = 10 \text{ ms}^{-2}$.

### (iii)   Combining the height and gravity standardisation

Suppose an athlete of height $h_1$ at a place of gravity $g_1$ has a maximum range $R_1$. The standardised range is calculated as follows.

Find $a$ from (6.2), then modify it for gravity with (6.4). Then from (6.3) the standardised maximum range is

$$R = \frac{g_1}{10} \left( \sqrt{h_1^2 + R_1^2} - h_1 \right)$$

a formula which is easy to apply by calculator or tables.

Although controversial, the notion of standardisation is familiar to other sporting activities such as weightlifting (Burghes *et al.*, 1982).

### INVESTIGATION

It is suggested that the reader might care to find out information about recent international field events and determine if the standardisation procedure described in this chapter will make appreciable differences to recorded ranges.

# 7

# Baiera's problem

## 7.1 THE PROBLEM AND BAIERA'S SOLUTION

In 1976, J. C. Baiera was a high school physics teacher in Pennsylvania. He noticed that schoolboys were competing in a national grading scheme for softball throwing in the playground. He had surveyed the playground and knew that it sloped upwards at a small angle around 2°, and realised that when the recorded distance for a boy was being measured along the incline, the boy was being cheated out of distance because the national averages were horizontal distances. The situation is illustrated in Fig. 7.1.

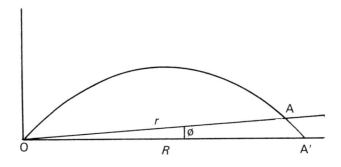

Fig. 7.1 — Baiera's problem.

The recorded distance is $OA = r$, but if there were no incline, the ball would land at A' and the range would then be $R > r$. In his paper, Baiera wished to correct the recorded distance $r$ and to reproduce the equivalent distance $R$ that would have been thrown along the horizontal, in order that his pupils were properly represented in the national scheme (Baiera, 1976).

He assumed that the ball was projected from ground level, at an angle of 45° to the horizontal, and that air resistance was negligible. If the speed of projection is $V_0$ then the equation of the trajectory is obtained from (2.9) as

$$y = x - \frac{gx^2}{V_0^2}$$

Referring to Fig. 7.1, the point A has coordinates $(r\cos\phi, r\sin\phi)$ and lies on the trajectory so that

$$r\sin\phi = r\cos\phi - \frac{gr^2\cos^2\phi}{V_0^2}$$

from which

$$\frac{V_0^2}{g} = \frac{r\cos^2\phi}{\cos\phi - \sin\phi}$$

Now if the playground were horizontal the maximum range would be $R = V_0^2/g$ so that

$$R = \frac{r\cos^2\phi}{\cos\phi - \sin\phi}$$

This formula enables the correction to be made. If a boy has a recorded distance of 60 m, and $\phi = 2°$, then the corrected distance would be $R = 62.13$ m, a difference of some 3.6%. Such a correction enabled many students to be moved into a higher percentile in the national standards.

Note that in his calculation, Baiera assumed that the ball would travel on the same trajectory whether or not the ground was level. In the next section we make use of the enveloping parabola to argue that this is not the case.

### 7.2 MODIFICATION AND SOLUTION OF BAIERA'S PROBLEM USING THE ENVELOPING PARABOLA

Baiera assumed that the softball would move on the same trajectory whether or not the ground was level. It is central to the modification detailed here that the boys had been practising in the same playground. The slope was too small to notice, but the boys who took the competition seriously would have discovered 'their' optimal angle of projection. The situation is illustrated in Fig. 7.2.

Clearly the trajectories will not be the same. In the case of the slope, the ball strikes the ground at A with coordinates $(r\cos\phi, -h + r\sin\phi)$ where $r = O'A$ along the slope. 'A' must be on the enveloping parabola, and so from (3.5) we obtain

$$2a(-h + r\sin\phi) = a^2 - r^2\cos^2\phi$$

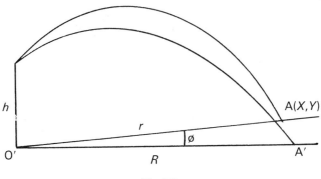

Fig. 7.2.

We use this equation to calculate $a$, a measure of the projection speed. We have

$$a^2 - 2(-h + r\sin\phi)a - r^2\cos^2\phi = 0$$

so that

$$a = -h + r\sin\phi + \sqrt{(h^2 - 2rh\sin\phi + r^2)}$$

(we can ignore the negative root).

Thus for any individual boy, we know $h$, $r$ and $\phi$, and hence we can calculate $a$. If the plane were horizontal then once again assuming the boy throws to achieve his maximum range, the ball hits the ground at $A'(R, -h)$. $A'$ is on the enveloping parabola and from (3.5) we obtain

$$R = \sqrt{(a(a + 2h))}$$

Thus having calculated $a$ we can obtain $R$, the corrected range. Unfortunately it is not possible to obtain a simple formula for the ratio $R$:$r$ as Baiera did since it depends upon both $h$ and $\phi$. However, if we take $\phi$ to be 2°, and $r$ as 60 m (a typical value from Baiera) and $h$ as 1.5 m (reasonable for a schoolboy) then $R/r = 1.034$. This is little different from Baiera's value because of the small values of $\phi$ and $h$.

In the Worked Example in Chapter 3 we found that the maximum range up a plane inclined at angle $\beta$, of a projectile thrown from a height $h$ above the ground, was

$$r_1 = \frac{-a\sin\beta + \sqrt{(a^2 + 2ah\cos^2\beta)}}{\cos^2\beta}$$

Consequently the range ($r_2$) down the plane is given by replacing $\beta$ by $-\beta$ in the above formula:

$$r_2 = \frac{a \sin \beta + \sqrt{(a^2 + 2ah \cos^2 \beta)}}{\cos^2 \beta}$$

Now it might be thought that a simple way to solve Baiera's problem would be to average $r_1$ and $r_2$. However, this proves not to be the case since

$$\frac{r_1 + r_2}{2} = \frac{\sqrt{(a^2 + 2ah \cos^2 \beta)}}{\cos^2 \beta}$$

which is not the maximum range on a horizontal plane. Note, however, that if $\beta$ is small and we employ the small-angle approximation then

$$\frac{r_1 + r_2}{2} = \sqrt{a^2 + 2ah}$$

which is the maximum range on the horizontal. So simply asking the boys to first throw up and then throw down the playground, and then averaging the two results will give the required range, provided $\beta$ is small, and provided the boys have been practising throwing both up and down in the playground. This last point is essential, since the two angles of projection will not be the same.

### 7.3  SOLUTION OF BAIERA'S PROBLEM WITH RESISTANCE PROPORTIONAL TO VELOCITY

In his solution, and our modification, air resistance has been neglected. The Reynolds number of the flow is large and so we shall assume that the resistance is proportional to the square of the velocity. Nevertheless, we first look at the case when it is proportional to $V$ because it is possible to make some progress since we have an analytic expression for the trajectory, and the solution is mathematically interesting.

As detailed in Chapter 4 this involves solving

$$X \tan \theta + \frac{gX \sec \theta}{kV_0} + \frac{g}{k^2} \log_e \left( 1 - \frac{kX \sec \theta}{V_0} \right) - Y = 0 \tag{7.1}$$

and

$$X = \frac{V_0^2}{kV_0 \sec \theta + g \tan \theta} \tag{7.2}$$

simultaneously for $V_0$ and $\theta$. We shall now write $k = k^{(1)}$ to emphasise that we are assuming $R \propto V$.

In Fig. 7.2, $A = (X, Y)$ and $X = r\cos\phi$, $Y = -h + r\sin\phi$ as before. To simplify matters we write $k^{(1)} = 0.1\,\text{s}^{-1}$,† $g = 10\,\text{ms}^{-2}$, $\phi = 2^0$, $r = 60\,\text{m}$ and $h = 1.5\,\text{m}$. The two equations then become

$$f_1(\theta, V_0) = 59.963\tan\theta + \frac{5996.3\sec\theta}{V_0} + 1000\log_e\left(1 - \frac{5.9963\sec\theta}{V_0}\right) - 0.59397 = 0$$

(7.3)

and

$$f_2(\theta, V_0) = 5.9963 V_0 \sec\theta + 599.63\tan\theta - V_0^2 = 0 \tag{7.4}$$

Two simultaneous non-linear equations in two variables can be solved by Newton–Raphson iteration in the following way.

If $f_1(x) = 0$ and $f_2(x) = 0$ then the sequence of vectors, $x_n$, obtained from the iteration

$$x_{n+1} = x_n - J^{-1}(x_n)f(x_n)$$

converges (under certain conditions) to a root of the simultaneous equations, where

$$x = \begin{pmatrix} \theta \\ V_0 \end{pmatrix} \qquad x_n = \begin{pmatrix} \theta \\ V_0 \end{pmatrix}_n$$

$$f(x_n) = \begin{pmatrix} f_1(x_n) \\ f_2(x_n) \end{pmatrix}$$

$J$ is the Jacobian matrix of $f$, i.e.

$$J = \begin{pmatrix} \dfrac{\partial f_1}{\partial \theta} & \dfrac{\partial f_1}{\partial V_0} \\ \dfrac{\partial f_2}{\partial \theta} & \dfrac{\partial f_2}{\partial V_0} \end{pmatrix}$$

and $J^{-1}$ its inverse.

Since $J$ is a $2 \times 2$ matrix, its inverse can be written down explicitly. Computer Program 4 in Appendix III carries out such an iteration. For starting values of $\theta = 40^0$ and $V_0 = 24\,\text{ms}^{-1}$, six iterations produce an answer of $\theta = 41.990^0$, $V_0 = 27.613\,\text{ms}^{-1}$ to 3 decimal places. We conclude that under such conditions the softball

† The justification for this value is given in the next section.

must have been thrown at $27.613\,\mathrm{ms}^{-1}$ to achieve this maximum range of 60 m on the inclined plane. We now use this to find the maximum range on the level. Equations (7.1) and (7.2) become

$$R \tan \theta + 3.6215R \sec \theta + 1000 \log_e (1 - 0.0036215R \sec \theta) + 1.5 = 0 \quad (7.5)$$

and

$$R - \frac{762.478}{2.7613 \sec \theta + 10 \tan \theta} = 0 \qquad\qquad (7.6)$$

with $V_0 = 27.613$, $Y = -1.5$, and all the other parameters having their previous values.

The same technique of solution applies. From an initial estimate of $R = 60$ m and $\theta = 40^0$, 3 iterations produce answers of $R = 61.812$ m and $\theta = 40.955^0$, to 3 decimal places. We conclude that a reasonable estimate of $R$ is $61.812$ m, i.e. $R/r = 1.03$.


## 7.4  SOLUTION OF BAIERA'S PROBLEM WITH $R \propto V^2$

Since we have no analytic expression for the equation of trajectories, the problem is now more complex. We have assumed that the magnitude of the resistive force, $R$, is given by $R = mkV^2$. From Chapter 5 we can say that $k - \rho C_\mathrm{D} A/2m$. The relevant parameters for a typical softball are $m = 0.17\,\mathrm{kg}$, $C_\mathrm{D} = 0.2$, $A = 0.00724\,\mathrm{m}^2$ and $\rho = 1.22\,\mathrm{kgm}^{-3}$, so that $k = k^{(2)} = 0.005$. (We have written $k = k^{(2)}$ to emphasise that this value of $k$ is associated with $R \propto V^2$.) At a typical speed of $25\,\mathrm{ms}^{-1}$

(i)  in the case of $R \propto V^2$, $R = mk^{(2)}V^2$
$$= 0.53\,\mathrm{N}$$

(ii)  in the case of $R \propto V$ used as an approximation to (i) we wish $R$ to have the same value. Therefore

$$0.53 = mk^{(1)}V$$
$$\Rightarrow k^{(1)} = 0.12$$

Earlier in examples and exercises we took the value of $k^{(1)}$ to be 0.1 for ease of arithmetic.

Here we describe a numerical approach to the solution. Suppose we know that a particular boy has achieved a range of 60 m on the inclined plane. We know the coordinates of $A(X, Y)$ in Fig. 7.3. (We shall assume that $\phi = 2^0$, $h = 1.5$ m and $k = 0.005$.) We now select a (fairly low) value of $V_0$ and solve the equations of motion

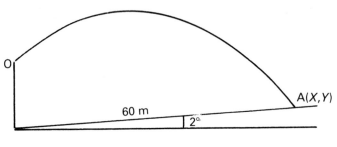

O

60 m

2°

A(X,Y)

Fig. 7.3.

numerically for a range of values of θ. Computer Program 5a in Appendix III does this, and the trajectories are illustrated in Fig. 7.4.

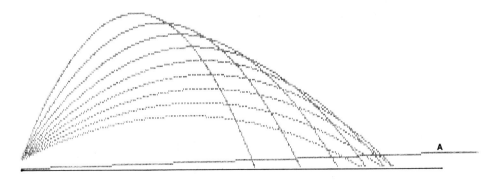

A

Fig. 7.4 — Trajectories and enveloping curve for $R \propto V^2$.

It is evident from the diagram that an enveloping curve exists even though we have not obtained its mathematical formula. Clearly no trajectory reaches the point A. Therefore the ball must have been thrown with a greater initial velocity. The process is repeated with $V_0$ increased by a small amount. Eventually, perhaps after many repetitions, a velocity $V_M$ is reached for which a trajectory just passes through A. This is the minimum velocity that is required to take the softball to the point A. Only larger values of $V_0$ will have trajectories which also pass through A. $V_M$ is the value we seek since we have assumed that the boy has done his best and has been practising on the inclined plane, and so cannot have thrown faster than $V_M$. Had he done so, the ball would have travelled more than 60 m.

Once we know $V_M$ we now solve the equations of motion with $V_0 = V_M$ for a range of values of θ. The maximum range on the ground obtained in this way gives the required correction. If this procedure is carried out using Computer Program 5b we obtain $V_M = 27.63\,\mathrm{ms}^{-1}$ and the corrected range, $R$, is 60.6 m. We conclude that

$R/r = 1.01$. We note that the ratio 1.01 is somewhat smaller than Baiera's value of 1.036 and so he was probably overcompensating his pupils.

## EXERCISES

1. Rework the problem using different parameters (e.g. change $h$). How sensitive is the quantity $R/r$ to such changes? If one value of $R/r$, say 1.01, were used for all boys, would the taller or the shorter ones be advantaged? Investigate the changing of other parameters.
2. Recall that in the absence of air resistance the average of the ranges up and down the playground gave a good approximation to the range on the horizontal, when the slope of the playground was such that the small-angle approximations could be used. Investigate numerically whether or not such an approximation is reasonable in the case of resisted motion.
3. Re-examine this problem using the equation of the envelope directly. (The envelope for the case $R \propto V$ is given in Chapter 4.)

# 8

# The Percy Grainger problem

## 8.1 INTRODUCTION

Percy Aldridge Grainger, born in July 1882, was one of Australia's best-known composers. He loved sport — running was his main physical recreation — and he soon become known as the jogging pianist! (Bird, 1976). He is reputed to have been able to throw a cricket ball from one side of his house to the other, and catch it descending on the other side by running through the house. In this chapter we look a little more closely at this problem and its possible solutions.

## 8.2 THE MODEL AND A SOLUTION

We will assume that the cricket ball is thrown with speed $V_0$ over the symmetrical house illustrated in Fig. 8.1. The thrower can run straight through the house through

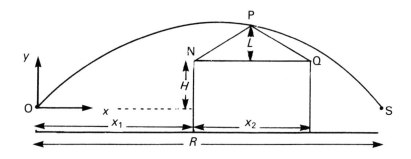

Fig. 8.1— The Percy Grainger problem.

open doors without hindrance.

The problem, given $L, H, x_1$, and $x_2$, is to find the possible angles of projection, $\theta$. If $\theta$ is too small, the ball will hit the side of the house or the nearside roof. If $\theta$ is too large, the ball may not clear the roof. If the thrower runs too slowly, he may not arrive in time to catch the ball, although we are not worried if he arrives too early and has to wait until it finishes its descent.

From the standard projectile formula given in Chapter 2 we know that in the absence of air resistance, the time of flight $= 2V_0\sin\theta/g$, the horizontal range, $R$, is $V_0^2\sin2\theta/g$ and the trajectory is given by

$$y = x\tan\theta - \frac{gx^2}{2V_0^2}(1 + \tan^2\theta)$$

Let the thrower be *stationary* when the ball is released. We will assume that the ball is caught at the height of release, i.e. when $x = R$. The horizontal speed of the ball is always $V_0\cos\theta$. This must be the minimum average speed of the thrower in order that he can arrive at the point S 'in time'. The quantities in Fig. 8.1 are required to satisfy the following inequalities.

(1) When $x = x_1 + x_2/2$, $y > L + H$.

This means the ball clears point P.

(2) When $x = x_1 + x_2$, $y > H$.

This means the ball clears point Q.

(3) When $x = x_1$, $y > H$.

This means the ball clears point N. The three conditions together mean the ball clears the house.

We choose particular values of $L$, $H$, $x_1$, $x_2$ and $V_0$, as these would be given in any situation. Take $H = 5$, $L = 2$, $x_1 = 10$, $x_2 = 10$, $g = 10$ and $V_0 = 20$ in appropriate units.

Imposing the first requirement yields a quadratic inequality for $T = \tan\theta$:

$$2.8125T^2 - 15T + 9.8125 < 0$$

the solution of which shows that $0.763 < T < 4.57$ so that $37.34° < \theta < 77.66°$. This is the condition required in order that the cricket ball clears the top of the roof. The reader can verify that condition (3) is automatically satisfied when $\theta$ lies in this range.

Imposing the second requirement yields another inequality in $T = \tan\theta$:

$$T^2 - 4T + 2 < 0$$

with solution $0.586 < T < 3.414$ so that $30.36° < \theta < 73.68°$. Combining these two inequalities gives

$$37.34° < \theta < 73.68°$$

Now the thrower must run at a speed greater than or equal to $V_0\cos\theta$. Since $\cos\theta$ is a decreasing function of $\theta$ in the range $0° < \theta < 90°$, the minimum speed, as a function of $\theta$, occurs when $\theta$ is maximum, i.e. $73.68°$. The average running speed of the thrower must be greater than $20\cos73.68° = 5.62\,\text{ms}^{-1}$ or $12.6\,\text{mph}$. When $\theta = 45°$ the average running speed of the thrower is $14.14\,\text{ms}^{-1}$ or $31.6\,\text{mph}$. Obviously, running speeds as large as this are impossible, so that although it may be easier to throw the ball at $45°$ and still clear the roof, there will be no chance of catching it as it descends.

## EXERCISES

1. The angle of $73°$ obtained above is rather too large for practical purposes; it would be difficult to achieve a reasonable projection speed at such an angle. Given that the average running speed is $10\,\text{ms}^{-1}$, what angle should the ball be thrown at? Is a practical solution feasible?

2. Given that the size of the house is fixed, and the projection speed $V_0$ is $20\,\text{ms}^{-1}$, what variables can be changed? (For example, you could throw from further away.) Estimate some typical parameters, e.g. the size of your house, and repeat the above analysis.

3. For the same house, suppose, now, that $V_0$ can vary up to a maximum of $30\,\text{ms}^{-1}$. Investigate the effects of variation of speed and distance from the house when the average running speed is no greater than $10\,\text{ms}^{-1}$. Remember the ball must clear the house and you must be able to run fast enough to catch it.

4. What is the size of the largest house for which the problem has a solution? Let the throwing speed not exceed $30\,\text{ms}^{-1}$ and the running speed not exceed $10\,\text{ms}^{-1}$.

# 9

# Safe driving speeds on newly surfaced roads

## 9.1  THE PROBLEM

Cheshire County Council in common with many local authorities recommends that drivers limit their speeds to 20 mph when travelling over roads which have been newly dressed with a bituminous binder and stone chippings. One unavoidable snag with this method of maintenance is that some surplus stones remain on the road until they are swept up some hours after laying. The advisory limit is intended to avoid the situation where stones are caused to fly into the path of other vehicles and so cause damage to paintwork and even breakage of windscreens. In this chapter we present various mathematical models which describe some aspects of this situation, and we argue that a 'safe' speed is not significantly lower than that recommended.

## 9.2  ASSUMPTIONS

1. The wheel does not skid. Owing to a variety of factors such as the nature of the tyre, the tackiness of the binder, the weight and size of the vehicle, stones are thrown up at random speeds not greater than the speed of the vehicle. We therefore assume the simplest case in which stones are projected with the speed of the vehicle.
2. In practice it is observed that the direction of projection is random, but in the first simple model we assume that a stone remains in the vertical plane containing the direction of motion of the vehicle.
3. Vehicles travel at constant speed, their separation from the vehicle in front being not less than that recommended by the Highway Code.
4. Air resistance is negligible.
5. Drivers are not unduly worried by stones hitting the front bumper, grille, etc.

In the solution we shall make use of the bounding parabola.

## 9.3   SOLUTION

A stone is projected from the origin 0 at time $t=0$ and moves along a typical trajectory as shown in Fig. 9.1. The front of the bonnet of the following car, FC, has

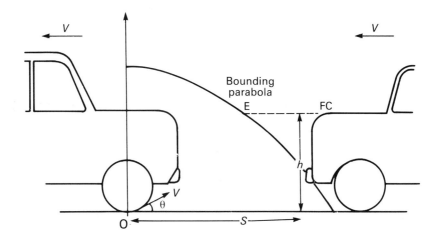

Fig. 9.1.

coordinates $(S,h)$. As a first attempt we will argue as follows.

There is certainly a danger of the bonnet, windscreen or top of the car being hit by the stone if FC has crossed the bounding parabola, i.e. if FC is within the bounding parabola at time $t_E$, the time of flight of the stone from O to E.

After time $t_E$ the position of FC is

$$S - Vt_E$$

and so the vehicle is in danger of being hit if

$$S - Vt_E < X_E \tag{9.1}$$

where $X_E$ is the $x$ coordinate of E.


## 9.4   SEPARATION DISTANCES

An expression for $S$ is now required and experience shows this to be related to the speed of travel $V$. We will assume that drivers adhere to the Highway Code recommended separation distances. This assumption enables us to determine $S$ as a function of $V$ in the following way.

The Highway Code provides the information depicted in Fig. 9.2. The data show that $S$ can be divided into two parts, namely (a) the thinking distance $S_T$ (which is the distance travelled in the time it takes for the driver to react and depress the brake pedal), and (b) the braking distance $S_B$ (which is the distance travelled whilst braking from speed $V$ to rest).

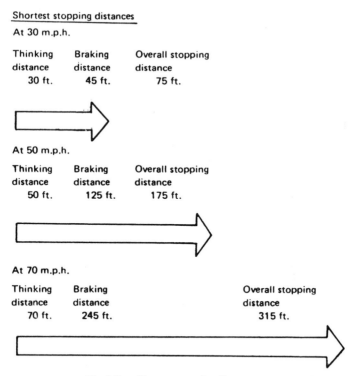

Shortest stopping distances

At 30 m.p.h.

| Thinking distance | Braking distance | Overall stopping distance |
|---|---|---|
| 30 ft. | 45 ft. | 75 ft. |

At 50 m.p.h.

| Thinking distance | Braking distance | Overall stopping distance |
|---|---|---|
| 50 ft. | 125 ft. | 175 ft. |

At 70 m.p.h.

| Thinking distance | Braking distance | Overall stopping distance |
|---|---|---|
| 70 ft. | 245 ft. | 315 ft. |

Fig. 9.2 — Shortest stopping distances.

It is clear from the data that $S_T = V$ feet.

To obtain the formula between $S_B$ and $V$ it is useful to assume a relationship of the form $S_B = kV^\alpha$ where $k$ and $\alpha$ are constants. Then $\log S_B = \alpha \log V + \log k$.

Plotting $\log S_B$ against $\log V$ should yield a straight-line graph if the assumed relationship is true. From such a graph the values of $\alpha$ and $k$ can be determined. This graph is illustrated in Fig. 9.3.

The graph is indeed a straight line of gradient 2, so that the assumed relationship between $S_B$ and $V$ is true, and the value of $\alpha$ is 2. Furthermore the value of $k$ can be shown to be 0.05. We conclude that $S = V + 0.05V^2$, where $S$ is measured in feet and $V$ in mph. However, our model requires the use of SI units. It is straightforward to show that the equivalent formula when $S$ is measured in metres and $V$ in metres per second is

$$S = 0.682 + 0.076V^2 \tag{9.2}$$

## 9.5 SOLUTION CONTINUED

Since E lies on the bounding parabola its $x$ coordinate is obtained from (3.5) as

$$X_E = \sqrt{(a(a - 2h))} \tag{9.3}$$

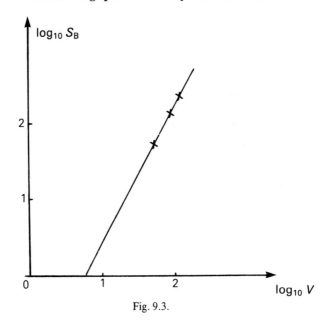

Fig. 9.3.

where $a = V^2/g$. Furthermore from (3.8)

$$\tan\theta = a/x \qquad (9.4)$$

whence $\cos\theta = x/\sqrt{(a^2 + x^2)}$ and since $x = Vt\cos\theta$ we obtain

$$t_E = \frac{\sqrt{(a^2 + X_E^2)}}{V} \qquad (9.5)$$

Combining (9.1), (9.3) and (9.5), we have the result that the following vehicle is in danger of being hit by the stone if

$$S - \sqrt{(2a(a - h))} < \sqrt{(a(a - 2h))} \qquad (9.6)$$

Therefore from (9.2), there is danger if

$$\sqrt{g}(0.682 + 0.076V) < \sqrt{(V^2/g - 2h)} + \sqrt{\left(2\left(\frac{V^2}{g} - h\right)\right)} \qquad (9.7)$$

## 9.6 INTERPRETATION OF THE RESULT

If the equality given by (9.7) is solved numerically for $h = 0.75$ m (typical of many popular makes of car) we find that if a driver exceeds 5.531 ms$^{-1}$ (12.372 mph) he is

in danger of being hit. It might be argued that even though the car remains outside
the bounding parabola at time $t_E$, at some later time when it is within the bounding
parabola it could still be hit by a stone which has taken time $t_h > t_E$ to reach height $h$.
It can easily be shown that the time taken for a stone to rise and then fall to height $h$ is

$$t_h = \frac{V\sin\theta + \sqrt{(V^2\sin^2\theta - 2gh)}}{g} \tag{9.8}$$

from which we deduce that $t_h$ is an increasing function of $\theta$. Taking $V = 5.531\ \mathrm{ms}^{-1}$
and $h = 0.75\ \mathrm{m}$, equations (9.3) and (9.4) imply that $\theta_E$, the angle of projection which
takes a stone to the point E on the bounding parabola, is about 54°. Consequently a
greater time to reach a height $h$ must arise from any angle of projection greater than
this. However, observation shows that the wheel arches and rubber flaps of the
vehicle constrain angles to be much less than 54°. We therefore conclude that if
vehicles travel at $5.531\ \mathrm{ms}^{-1}$ with a separation of at least that recommended by the
Highway Code, there is no danger of being hit. The value of $5.531\ \mathrm{ms}^{-1}$ obtained in
this way is rather lower than the advisory limit recommended by local authorities,
and in practice, drivers are reluctant to slow down even to 20 mph and would almost
certainly ignore requests to slow down even further.

## 9.7   MODEL 2

Suppose, now, that we no longer assume a Highway Code separation. Consider a
stone projected from O (Fig. 9.1). The time it takes to reach height $h$ as shown is
given by (9.8). Its $x$ coordinate is then

$$x_h = V\cos\theta t_h \tag{9.9}$$

If the following car has separation $S_1$ given by

$$S_1 = x_h + Vt_h \tag{9.10}$$

then a stone projected at angle $\theta$ will reach the bonnet at time $t_h$. Our procedure is as
follows: choose $V$ and calculate the smallest angle $\theta_{MIN}$ that can cause a stone to
reach height $h$ at this speed, then gradually increase $\theta$ from this smallest value up to
$\pi/2$, at each stage calculating $S_1$ from (9.10). Simultaneously we calculate
$S = 0.682V + 0.076V^2$, the Highway Code separation. We then gradually increase $V$
and repeat the whole procedure. It soon becomes apparent that at very low speeds,
$S_1 < S$ for all angles of projection. This means that if a vehicle travels at the
recommended separation it must be outside the range of the furthest stone to reach
height $h$. Eventually, as $V$ is increased there comes a point when $S_1 > S$ for some
angles of projection, meaning that a vehicle travelling at the recommended separation can be hit on the front of the bonnet by a stone at some angle. This searching
problem is easily handled on a microcomputer. For $h = 0.75\ \mathrm{m}$ it reveals that if $V$ is
less than about $5.17\ \mathrm{ms}^{-1}$, a car travelling at the recommended separation is safe.

Above this speed there is the possibility of a stone hitting the car. Again a value of $5.17\,\text{ms}^{-1}$ is unrealistic in practice, but we can argue that instead of allowing $\theta$ to vary from $\theta_{\text{MIN}}$ to $\pi/2$, we constrain it to lie between $\theta_{\text{MIN}}$ and 35°, say. The searching routine so modified yields a speed of approximately $8\,\text{ms}^{-1}$ or 18 mph. Measurement of the overhang of the boot on many makes of car leads us to the conclusion that 35° is a maximum value, stones projected at a higher angle hitting the underside of the vehicle. Many saloon-type cars have a much smaller maximum. (We are of course neglecting tractors, etc.) This second model leads us to the conclusion that for most types of vehicle travelling at the Highway Code recommended separation, a safe driving speed is not significantly lower than that recommended by the local authority.

### EXERCISES AND FURTHER INVESTIGATIONS

1. Trace the path of a typical stone launched with a typical velocity, to find the point where it hits the windscreen of a following car. Measure the inclination of the windscreen. Calculate the components of the stone's velocity both parallel and perpendicular to the windscreen at the time of impact. Calculate the impulse on the windscreen (you will need to estimate the mass of a stone chipping). It may be possible for you to obtain data on how large an impulse a windscreen can stand.

2. In the previous analysis, motion was restricted to the vertical plane which contained the direction of motion of the car. Now consider the general three-dimensional case.
    (i) Write down and solve the equations of motion for a stone projected from the origin with speed $V$ at angle $\theta$ to the horizontal and at an angle $\phi$ from the $x$ axis as shown in Fig. 9.4.
    (ii) Obtain the equation of the trajectory of the stone.

3. As in the two-dimensional case, certain regions are accessible and others are not. In the general case, the boundary between the two regions is known as the enveloping paraboloid. Investigate ways of obtaining its equation.

4. A stone is thrown up from the tyre of a car travelling at speed $V$. Its initial direction of motion makes an angle $\theta$ with the horizontal and an angle $\phi$ with the $x$ axis. Investigate whether a car travelling in the same direction directly behind the first car is in danger of being hit.

5. Extend the investigation of the last problem to the case when the following vehicle is in an overtaking position but still travelling with the same speed. The Highway Code recommends that vehicles should leave a gap of a 'car door's width' when overtaking. This gap is approximately 1 m.

6. A motor car is travelling at a constant speed $V$. Lumps of mud are being thrown tangentially from points of the back tyres where the tangent makes an angle $\Phi$ or less with the road. The radius of the tyres is $r$. Investigate whether a car following at the same speed will have its windscreen sprayed with mud if the lowest point of the windscreen is a height $h$ above the road. Assume the following car is a distance

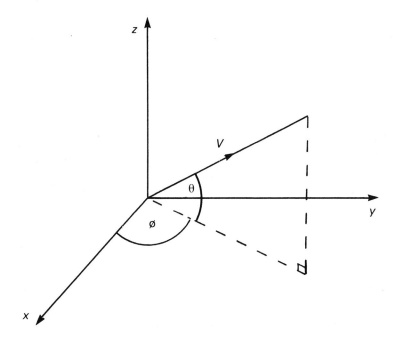

Fig. 9.4.

*D* behind the first. Measure suitable values of *D* and *r* and try to estimate a reasonable value of Φ.

7. Extend the problem of stone chippings being thrown up from a newly resurfaced road to the case where two vehicles are moving in opposite directions.

# 10

# Modelling the flight of a javelin

## 10.1 INTRODUCTION

For many thousands of years, man has exploited the power of a projectile in motion to gain food by hunting with spears and arrows, and to gain advantage on the battlefield with a variety of devices. One such device was the javelin, and in this chapter we model some aspects of its flight. Hunting and warring aside, the early Greeks and Romans enjoyed javelin throwing as a competitive sport. The traditional date for the start of the Olympic Games is generally accepted as 776 BC, although this was not the beginning of Greek athletics. As their empire grew, the Greeks scattered to all parts of the Mediterranean, periodically gathering at a number of sites for religious festivals. At these festivals a variety of sports were played. Because various sites were used, some measure of standardisation became necessary and so the Olympic Games came into being. At the same time the javelin retained close links with training for war, because it was a vital component of the armament of an infantryman who usually carried two. Infantrymen launched javelins at the enemy before coming close enough to use swords. Consequently, accuracy and range were very important — a well-thrown javelin secured an immediate advantage over the other side. There is no record of the distances thrown by the ancient Greeks, so we have no means of knowing what they achieved compared to our modern-day athletes, the best of whom throw to distances of over 80 m for men and 70 m for women (Fig. 10.1).

## 10.2 THE JAVELIN AS AN AERODYNAMIC DEVICE

The mass of a typical javelin is 0.8 kg. If all this mass were concentrated at one point and this point behaved like the shot put in Chapter 6, thrown at a typical speed of $30 \text{ ms}^{-1}$ from ground level, its maximum range, in the absence of air resistance, would be

$$R = \frac{V_0^2}{g} = 90 \text{ m}$$

Fig. 10.1

from (3.9) if $g = 10 \text{ ms}^{-2}$.

A sphere of the same mass, thrown in the atmosphere, has a much smaller range because of the resistance of the atmosphere. The larger the radius of the sphere, the more the resistance, and so the greater the reduction in the range. However, observation of javelin throws in world record field events shows that ranges of 80–90 m are typical, and so the javelin is obviously a much more complicated aerodynamic device. The reason for this increase in range lies in its aerodynamic behaviour — its elongated shape allows lift to be generated which causes an increase over and above the loss due to frictional drag.

In the case of the shot put, the aerodynamic force is negligible compared with the weight. However, in order to model the flight of a javelin successfully we must take the aerodynamic force into account.

## 10.3  ASSUMPTIONS OF THE MODEL

(i)  We consider the javelin to be a rod which remains in the vertical plane of projection. This means that it is allowed to 'pitch', i.e. rotate in this vertical plane, but not 'yaw', so that lateral motion does not occur (Fig. 10.2).

(ii)  We will assume that the javelin does not spin about its own axis. In actual fact a good javelin thrower will impart spin to the javelin when it is released from the hand. This has the effect of stabilising the flight.

(iii)  The influence of the wind is negligible.

(iv)  We will assume that the coefficients of lift and drag depend only upon the angle of attack, and that the lift and drag forces are proportional to the square of the speed.

The justification for these assumptions is now given: the drag and lift are complex quantities, and at best only experimental results are available. The data used in this model are appropriate for the 'TI APOLLO' (new rules) javelin. Best & Bartlett (1987) have found that both the lift and the drag vary as the square of the velocity. Recall from Chapter 5 that $L = \frac{1}{2}\rho C_L A V^2$ and $D = \frac{1}{2}\rho C_D A V^2$, so that in this model the coefficients of lift and drag are independent of velocity. To see this, consider Fig. 10.3(a) which shows the experimentally determined lift and drag against $V^2$, at a fixed angle of attack of 30°. It is clear that the dependence upon $V^2$ is linear, i.e. that the dependence upon $V$ is quadratic.

The gradient of these graphs will give the factors $\frac{1}{2}\rho C_L A$ and $\frac{1}{2}\rho C_D A$ at that angle of attack, which for convenience we will denote by $K_L$ and $K_D$ respectively. Figs 10.3(b,c) show the results of measuring these factors at various angles of attack. Best & Bartlett (1987) fitted suitable approximating curves to these data in order to predict $K_D$ and $K_L$ at any $\alpha$. Following them we will assume that

$$K_D = 0.00024e^{0.09\alpha} \tag{10.1}$$

$$K_L = 0.0000561\alpha^{1.34}$$

We are now in a position to calculate the drag and lift forces once we know the speed and angle of attack.

(v)  Recall from Chapter 5, that although the lift and drag arise from pressure and frictional forces acting over the whole surface of the body, it is mathematically convenient to consider them to be acting at a single point called the *centre of pressure*. The centre of pressure can be determined experimentally in a wind tunnel by carefully balancing the javelin. Some workers have found that as the angle of attack varies, the centre of pressure moves from one side of the centre of gravity to the other. However, for the javelin considered in this model, the centre of pressure was found to be at a fixed distance behind the centre of gravity. This is fortunate, as it makes the mathematical analysis much more tractable.

## 10.4  NOTATION

In Fig. 10.4, G denotes the centre of gravity, P the centre of pressure, and **V** the velocity vector of the centre of gravity. The pitch is denoted by θ, α is the angle of

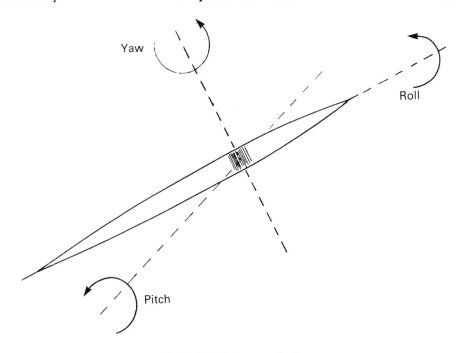

Fig. 10.2 — Pitch, yaw and roll.

attack and $\phi$ is the angle the velocity vector makes with the horizontal. It is necessary to introduce the angle $\phi$ because the direction of motion of the centre of mass is not the same as the inclination of the javelin to the horizontal. The drag opposes the motion, and so is in a direction opposite to **V**, and the lift is perpendicular to the drag. The weight of the javelin is $mg$. The centre of pressure is a distance $d$ behind the centre of gravity.

## 10.5   THE EQUATIONS OF MOTION

When a rigid body such as the javelin is in motion, it is conventional to consider separately the translational motion of the centre of gravity, and the rotational motion about this point (Chorlton, 1978). Taking $x$ and $y$ axes as shown it is now straightforward to write down the horizontal and vertical equations of motion.

Horizontally     $- D\cos\phi - L\sin\phi = m\ddot{x}$

Vertically     $- mg - D\sin\phi + L\cos\phi = m\ddot{y}$     (10.2)

For the rotational motion about G we have

$$(- D\sin\alpha - L\cos\alpha)d = I\ddot{\theta}$$

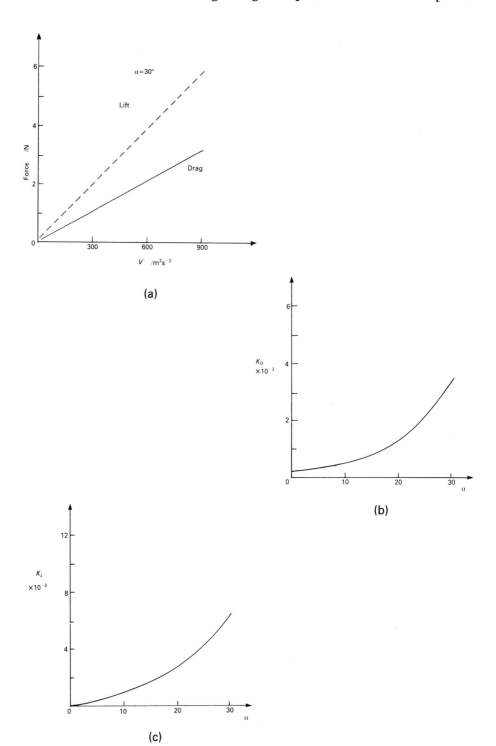

Fig. 10.3

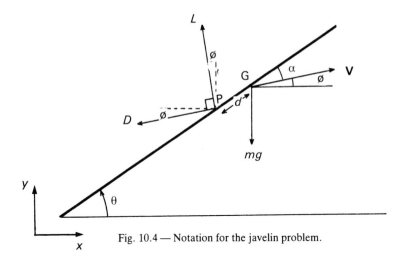

Fig. 10.4 — Notation for the javelin problem.

where $I$ is the moment of inertia of the javelin about the transverse axis through its centre of gravity, which must be experimentally determined. Note that $\phi = \tan^{-1}(\dot{y}/\dot{x})$ and $\theta = \alpha + \phi$, so that knowing $x(t)$, $y(t)$ and $\theta(t)$ the problem is solved since $D$ and $L$ are given at any speed and angle of attack from equation (10.1).

We therefore have three second-order linked ordinary differential equations to solve for $x, y$ and $\theta$.

## 10.6   DATA

Typical values of the physical constants are given below:

Mass $m = 0.80625$ kg
$\quad\quad d = 0.255$ m
$\quad\quad I = 0.42$ kg m$^2$

The distance from the centre of gravity $G$ to the tip of the javelin is 1.057 m.

## 10.7   SOLUTION OF THE EQUATIONS OF MOTION

It is impossible to solve the three second-order equations analytically and so we adopt a numerical approach. It is usual in such circumstances to write the three equations as an equivalent system of six first-order equations:

$$m \frac{du}{dt} = -D\cos\phi - L\sin\phi$$

$$m \frac{dw}{dt} = -mg - D\sin\phi + L\cos\phi$$

$$I \frac{d\omega}{dt} = (-D\sin\alpha - L\cos\alpha)d$$

$$\frac{dx}{dt} = u$$

$$\frac{dy}{dt} = w$$

$$\frac{d\theta}{dt} = \omega$$

Once a set of initial conditions is chosen, a unique solution of these equations can be determined. From this solution it is possible to find the range of the javelin. Since it is the aim of the thrower to maximise this range, the problem reduces to finding an optimal set of initial conditions. The above system of equations is solved by Computer Program 6 in Appendix III using the standard fourth-order Runge–Kutta method, which calculates the dependent variables at discrete time intervals $\Delta t = 0.05$ s, and plots the position and orientation of the javelin at time increments of 0.35 s. If the javelin lands tail-first ($\theta > 0$), a 'foul-landing' occurs. In the case of a successful landing, the program stops when the tip of the javelin touches the ground. Finally, the range is given. The program itself can be altered to vary physical constants such as the mass of the javelin. Some typical solutions are given in the next section.

## 10.8   SOME COMPUTED SOLUTIONS

Example 1 (Fig. 10.5).
Initial data:          $y = 2$ m                    $x = 0$ m
                       $\theta = 30°$               $\phi = 30°$
                       $V = 30\ ms^{-1}$            $\omega = 0°\ s^{-1}$

The landing was successful. The range $= 82.696$ m.

Example 2 (Fig. 10.6).
Initial data: as above but $\omega = 6°\ s^{-1}$.
   The user can see that imparting an initial angular velocity increases the range to 82.776 m.

Example 3 (Fig. 10.7).
Initial data:          $y = 2$ m                    $x = 0$ m
                       $\theta = 11.5°$             $\phi = 11.5°$
                       $V = 20\ ms^{-1}$            $\omega = 51.5°\ s^{-1}$

Fig. 10.5

Fig. 10.6

Because the landing was foul, in a competitive event this range would not be counted.

**EXERCISES**

1. For a fixed value of $V$, say 30 ms$^{-1}$, vary the initial pitch from about 20° to 50°. Take $\alpha = 0$. In each case, obtain the range. Try to locate which initial pitching

```
FOUL LANDING
RANGE=27.111
TIME OF FLIGHT =1.450
```

Fig. 10.7

angle gives the optimum range. Bear in mind that although this value may be the best dynamically, it may not be the best physically. The athlete may not be able to impart his optimum speed at such an angle.

2. Repeat the previous procedure with an increased value of $V$. How does this affect the optimum pitching angle?

3. Modify the program to obtain the value of lift throughout the flight. When is the lift negative?

# 11

# Throwing the cricket ball

## 11.1 INTRODUCTION

The cricket ball throw used to be a popular event at many school and amateur sports meetings. The athlete was expected to throw a cricket ball as far as possible. In order to increase the range, the athlete was allowed to throw the ball whilst running. For the purposes of this book the interesting mathematical model is expressed in the form of two problems and their solutions. The problems are different, but related. The reader should note that the model involves two frames of reference — the athlete's and that of a stationary observer.

## 11.2 PROBLEM No. 1

An athlete is observed to throw a cricket ball with speed $V_0$ at an angle $\theta$ to the horizontal, whilst running at a constant speed, $W$. The ball is released from a height $h$ above the ground.

(i) Find the angle, $\phi$, at which the athlete considers he has projected the ball, and the speed, $V$, at which he observes the ball to leave his hand.
(ii) Find $\theta$, $\phi$ and $V$ so that the ball is observed to have maximum range, if $V_0 = 20 \text{ ms}^{-1}$.

Solution:

(i) We use the Galilean Transformation (Alfonso & Finn, 1970). $x', u', v', y', t'$ are the projectile parameters with respect to the moving frame, and $x, u, v, y, t$ are those in the fixed frame, as illustrated in Fig. 11.1.

   The two coordinate systems coincide when $t' = t = 0$. The Galilean Transformation is

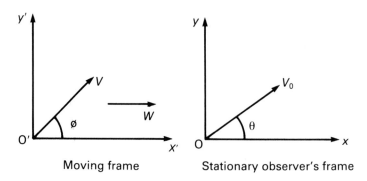

Fig. 11.1.

$$x' = x - Wt \qquad y' = y$$
$$u' = u - W \qquad v' = v$$

In the athlete's frame the speed of projection is given by

$$V = \sqrt{u'^2 + v'^2}$$
$$V = \sqrt{(u - W)^2 + v^2}$$
$$V = \sqrt{u^2 + v^2 - 2uW + W^2}$$

i.e.

$$V = \sqrt{V_0^2 - 2WV_0\cos\theta + W^2} \qquad (11.1)$$

Now similarly

$$\tan\phi = v'/u'$$
$$= v/(u - W)$$

i.e.

$$\tan\phi = \frac{V_0\sin\theta}{V_0\cos\theta - W} \qquad (11.2)$$

Numerical Example:

Take $\theta = 45°$, $V_0 = 20$ ms$^{-1}$, $W = 10$ ms$^{-1}$. To the athlete, the speed is 14.74 ms$^{-1}$ at an angle $\phi$ of 73.68° (from 11.1) and (11.2)).

Note that tan$\phi$ is infinite when $V_0\cos\theta - W = 0$ in (11.2). This merely means that the athlete considers that he has made a vertical throw.

(ii) From the assumptions in the question, $V_0$ is fixed and so this problem can be dealt with using the enveloping parabola as described in Chapter 3. Suppose $h = 2$ m. Then in the fixed frame, if the range is maximum we have (from 3.5))

$$2 \times \frac{400}{10} \times (-2) = 1600 - R^2$$

$$\Rightarrow \qquad R^2 = 1760$$

$$\Rightarrow \qquad R = 41.95 \text{ m}$$

Also $\tan\theta = a/x = 0.954$ so that $\theta = 43.64°$.

From (11.1), $V = 14.51$ ms$^{-1}$ and from (11.2), $\tan\phi = 3.09$ so that $\phi = 72.04°$.

## 11.3  PROBLEM NO. 2

We use the same notation as Fig. 11.1. An athlete throws a cricket ball at speed $V$ whether or not he is stationary. He runs at a constant speed $W$ and releases his ball at $t = 0$ at an angle $\phi$ to the horizontal, from a height $h$ above the ground. Find

(i) the speed and direction of the ball as seen by a stationary observer, and

(ii) $\phi$ to achieve maximum range.

Solution: From the Galilean Transformation we have

$$x = x' + Wt \qquad y = y'$$
$$u = u' + W \qquad v = v'$$

(i) In the fixed frame the speed of projection is given by

$$V_0 = \sqrt{u^2 + v^2}$$
$$V_0 = \sqrt{(u' + W)^2 + v^2}$$
$$= \sqrt{u'^2 + v'^2 + 2u'W + W^2}$$

i.e.

$$V_0 = \sqrt{V^2 + 2WV\cos\phi + W^2} \qquad\qquad (11.3)$$

Also $\tan\phi \ = \ v/u \ = \ v'/(u' + W)$

$$\Rightarrow \qquad \tan\theta \ = \ \frac{V\sin\phi}{V\cos\phi \ + \ W} \qquad\qquad\qquad (11.4)$$

Numerical example:

If $\phi = 45°$, $V = 20$ ms$^{-1}$ and $W = 10$ ms$^{-1}$, then for the stationary observer $V_0 \ = \ 27.98$ ms$^{-1}$ at angle of 30.36°, from (11.3) and (11.4).

(ii) Note now that $V$ is fixed, $W$ is constant and $\phi$ is variable. We want to find $V_0$ and $\theta$ so that the range is maximum. We cannot use the enveloping parabola in this case because $V_0$ is not constant since it depends upon $\phi$. Initially let us put $h = 0$. This is unrealistic but it does lead to an analytic solution. In the fixed frame the range, $R$, is given by

$$R \ = \ 2V_0^2 \sin\theta\cos\theta/g$$

We have
$$V_0\sin\theta \ = \ v \ = \ v' \ = \ V\sin\phi$$

and

$$V_0\cos\theta \ = \ u \ = \ u' + W \ = \ V\cos\phi + W$$

$$\Rightarrow \qquad R \ = \ \frac{2V(V\cos\phi \ + \ W)\sin\phi}{g} \qquad\qquad\qquad (11.5)$$

The reader should note that $V$ is fixed, unlike $V_0$. For maximum range we require $dR/d\phi = 0$:

$$\frac{dR}{d\phi} \ = \ \frac{2V}{g} \left( -V\sin^2\phi \ + \ (V\cos\phi + W)\cos\phi \right) \ = \ 0$$

$$\Rightarrow \qquad 2V\cos^2\phi \ + \ W\cos\phi \ - \ V \ = \ 0$$

so that

$$\cos\phi \ = \ \frac{-W \pm \sqrt{W^2 + 8V^2}}{4V}$$

Numerical example:

Let $W = 10$ ms$^{-1}$, $V = 20$ ms$^{-1}$ as before. The angle of projection in the moving frame for maximum range is $\phi = 53.62°$.

The corresponding values of $V_0$ and $\theta$ are given by (11.3) and (11.4). We have $V_0 = 27.15$ ms$^{-1}$ and $\theta = 36.37°$. The maximum range is given by (11.5) as 70.39 m.

Now suppose we let $h = 2$ m. In the fixed frame we have

$$x = V_0\cos\theta t \qquad y = V_0\sin\theta t - \tfrac{1}{2}gt^2$$

When $y = -2$ we have

$$-2 = V_0\sin\theta t - \tfrac{1}{2}gt^2$$

$\Rightarrow \qquad gt^2 - 2V_0\sin\theta t - 4 = 0$

$\Rightarrow \qquad t = \dfrac{2V_0\sin\theta \pm \sqrt{4V_0^2\sin^2\theta + 16g}}{2g}$

Since $\sqrt{4V_0^2\sin^2\theta + 16g} > 2V_0\sin\theta$ we take the positive root and obtain

$$t = \frac{V_0\sin\theta + \sqrt{V_0^2\sin^2\theta + 4g}}{g}$$

So the range, $R = V_0\cos\theta t$, is

$$R = \frac{V_0^2\sin\theta\cos\theta + V_0\cos\theta\sqrt{V_0^2\sin^2\theta + 4g}}{g}$$

Using the Galilean Transformation we have

$$R = \frac{V\sin\phi(V\cos\phi + W) + (V\cos\phi + W)\sqrt{V^2 \sin^2\phi + 4g}}{g} \qquad (11.6)$$

For maximum $R$ we require $dR/d\phi = 0$. However, this will not lead to an easy analytic solution and we therefore adopt a simple numerical approach as follows.

With $V = 20\,\text{ms}^{-1}$, $W = 10\,\text{ms}^{-1}$ we allow $\phi$ to vary from $0°$ to $90°$ in intervals of $5°$, and calculate $R$ from (11.6)

The results of this calculation are tabulated in Table 11.1, from which it is

**Table 11.1**

| $\phi°$ | Range (m) |
|---------|-----------|
| 0.000   | 18.974    |
| 5.000   | 24.847    |
| 10.000  | 31.740    |
| 15.000  | 39.138    |
| 20.000  | 46.521    |
| 25.000  | 53.465    |
| 30.000  | 59.647    |
| 35.000  | 64.826    |
| 40.000  | 68.830    |
| 45.000  | 71.543    |
| 50.000  | 72.900    |
| 55.000  | 72.884    |
| 60.000  | 71.519    |
| 65.000  | 68.872    |
| 70.000  | 65.043    |
| 75.000  | 60.168    |
| 80.000  | 54.408    |
| 85.000  | 47.944    |
| 90.000  | 40.976    |

clear that the optimum value of $\phi$ lies between $45°$ and $55°$. Table 11.2 shows the result of repeating the calculation for $45° < \phi < 55°$.

It is clear that the maximum range is approximately 73.06 m when $\phi = 52.5°$. This should be compared with the maximum range for a stationary throw: with speed $20\,\text{ms}^{-1}$ we have (from (3.5)) $R = 41.95$ m. Note that for maximum range, the apparent angle of projection to a stationary observer is $\theta = 35.6°$ and not $43.6°$ as might have (erroneously) been expected. (Why not?)

**Table 11.2**

| $\phi°$ | Range (m) |
|---|---|
| 45.000 | 71.543 |
| 45.500 | 71.740 |
| 46.000 | 71.924 |
| 46.500 | 72.094 |
| 47.000 | 72.251 |
| 47.500 | 72.393 |
| 48.000 | 72.522 |
| 48.500 | 72.637 |
| 49.000 | 72.739 |
| 49.500 | 72.827 |
| 50.000 | 72.900 |
| 50.500 | 72.960 |
| 51.000 | 73.007 |
| 51.500 | 73.039 |
| 52.000 | 73.058 |
| 52.500 | 73.063 |
| 53.000 | 73.055 |
| 53.500 | 73.032 |
| 54.000 | 72.997 |
| 54.500 | 72.947 |
| 55.000 | 72.884 |

**INVESTIGATIONS**

1. Extend Baiera's problem (Chapter 7) to the case where the thrower is running when he releases the softball.

2. In the light of the information given in this chapter, re-examine the Percy Grainger problem in Chapter 8.

# 12

## Harvesting grain

### 12.1 INTRODUCTION

In the prairies of North America and to a lesser extent in grain-growing areas of Britain, large mechanical harvesters are used. Fig. 12.2 shows such a harvester. Grain is projected from the blower (A) and is collected in the following trailer (B). The same situation occurs when forage is being collected. We will assume the grain or forage moves on a curve. The purpose of this chapter is to find an equation for that curve as seen by a stationary observer positioned some distance away.

### 12.2 THE MODEL

We make the following assumptions.

  (i)  The harvester moves with constant speed $U$.
 (ii)  Each particle of grain is thrown out from the harvester at a speed $V$ and at a fixed angle $\theta$, relative to the harvester.
(iii)  Air resistance is negligible.

We take axes as shown in Fig. 12.3.

    At time $t = 0$, the harvester is at O, and is moving with velocity $-U$. It projects grain continuously as it moves toward the left of the figure. Consider one particular particle of grain, thrown at time $t = \tau$, from the point with coordinates $(-U\tau, 0)$. Its initial speed relative to a fixed observer is $(V\cos\theta - U, V\sin\theta)$.

    The equation of motion is

$$\ddot{\mathbf{r}} = \begin{pmatrix} 0 \\ -g \end{pmatrix} \tag{12.1}$$

Fig. 12.1 — (Reproduced with permission from *Forage Conservation and Feeding*, Raymond Redmond and Waltham (Farming Press, Ipswich).)

Integrating and applying the initial condition we obtain

$$\dot{\mathbf{r}} = \begin{pmatrix} V\cos\theta - U \\ -g(t-\tau) + V\sin\theta \end{pmatrix} \qquad (12.2)$$

Similarly

$$\dot{\mathbf{r}} = \begin{pmatrix} (V\cos\theta - U)(t-\tau) - U\tau \\ \dfrac{-g}{2}(t^2 - \tau^2) + (V\sin\theta + g\tau)(t-\tau) \end{pmatrix} \qquad (12.3)$$

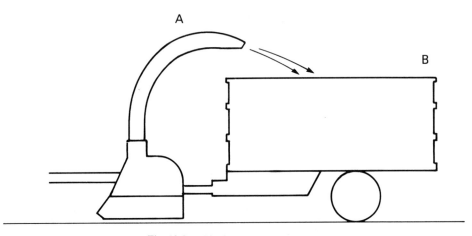

Fig. 12.2 — The harvester problem.

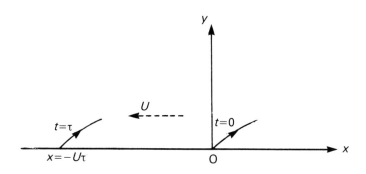

Fig. 12.3.

which can also be written as

$$\mathbf{r} = \begin{pmatrix} (V\cos\theta - U)(t - \tau) - U\tau \\ V\sin\theta(t - \tau) - \dfrac{g}{2}(t - \tau) \end{pmatrix}$$

Now as we consider different particles of grain, we are allowing $\tau$ to vary. As we let $\tau$ vary, (12.3) represents the parametric equation of the trajectory of each particle labelled by $\tau$, its time of release. (Note that $t$ is the parameter.)

Suppose at some fixed time $T > 0$ we take a photograph of the situation illustrated in Fig. 12.3.

What is the equation of the profile of the grain? This equation is found from (12.3) by fixing $t = T$ and eliminating $\tau$.

We have

$$x = (V\cos\theta - U)(T - \tau) - U\tau$$

$$y = V\sin\theta(T - \tau) - \frac{g}{2}(T - \tau)^2$$

Obtaining an expression for $\tau$ from the first of these equations and substituting in the second gives the required profile:

$$y = \left(\frac{x + UT}{V\cos\theta}\right)\left(V\sin\theta - \frac{g}{2}\left(\frac{x + UT}{V\cos\theta}\right)\right) \qquad (12.4)$$

(12.4) is the equation of the profile of the grain seen on the photograph.

Notes:
 (i) If $U = 0$ (i.e. the harvester is still and all particles move on the same trajectory) then

$$y = x\tan\theta - \frac{gx^2\sec^2\theta}{2V^2}$$

The reader should note that this is the trajectory equation first obtained in Chapter 2.
(ii) Rewriting (12.4) as

$$y = (x + UT)\tan\theta - \frac{g(x + UT)^2\sec^2\theta}{2V^2} \qquad (12.5)$$

we see that the equation of the profile has the form of the standard trajectory equation. In fact it is the trajectory of a fictitious particle fired from the harvester with speed $V$, at the time the photograph was taken but with the harvester stationary at the point with $x$ coordinate $-UT$. This is an interesting and unexpected result.
(iii) From (12.5) we have

$$\frac{dy}{dx} = \tan\theta - \frac{g\sec^2\theta(x + UT)}{V^2}$$

When $x = -UT$, $dy/dx = \tan\theta$. Therefore the gradient of the profile at its origin is the same as the tangent of the angle of projection relative to the harvester.

### 12.3  EXERCISES

1. Assuming that a jet of water from a water cannon behaves in the same way as the flow of grain described above, can you tell from a photograph whether or not the cannon is moving towards or away from the rioters, given that in either case it is moving at a constant speed. (You may assume that $U \approx 5\,\text{ms}^{-1}$, $V \approx 100\,\text{ms}^{-1}$ and $\theta \approx 10^0$.)

2. Instead of travelling at a constant speed $U$, the harvester accelerates from rest at a constant rate $f\,\text{ms}^{-2}$. Obtain the equation of the profile seen in a photograph taken at some later time $T$ in terms of the parameter $\tau$.

3. During a recent riot in Ruritania the authorities attempted to control the mob by means of a water cannon. Lawyers representing the rioters later claimed that the police action was provocative since the cannon was advancing towards the crowd. The police denied this, claiming they were in fact retreating. Both sides agree that the vehicle was acclerating at $1\,\text{ms}^{-2}$ and that the water was ejected at $100\,\text{ms}^{-1}$ at an angle of $10^0$.

    Is it possible to decide between the two sides on the basis of a press photograph?

# 13

# Modelling the flight of a golf ball

## 13.1  INTRODUCTION

The Romans played a game, *Paganica*, in which a leather-covered ball filled with feathers was hit with a bent stick. This, it is argued, was one of the origins of golf, but it was the fifteenth or sixteenth century before the ball was not only hit a long distance but was also manoeuvred into a small hole.

The standards of golf depend upon the type of equipment used. One of the earliest golf records was recorded on a winter's day in 1836 when a Frenchman, Messieux, a teacher at Madras College at St Andrews, drove a feathery ball 361 yards. The ball used was a Gourlay ball, the best of the day, which cost five shillings. The best of the feathery balls were always consistent in size, shape and surface. The stitched seams, which some golfers felt were a hindrance, improved the ball's aerodynamics.

In the mid-nineteenth century the gutta-percha ball (the 'guttie') was invented by the Reverend Dr Patterson. He received some gutta-percha as packing for a parcel sent from India, and shaped the latex-like substance into a ball and used it to play golf. The early attempts were unsuccessful mainly because the ball was smooth and had poor aerodynamic qualities and dipped sharply back to earth after travelling only a short distance. After some use, however, the ball became scratched and its performance improved. The ball makers soon discovered the trick of hammering indentations on the balls. This was soon mechanised and moulds were devised to impart dimples or 'blackberry' markings on the surface of the ball.

At the turn of the twentieth century an Ohio dentist, Dr Haskell, invented the rubber-cored ball. Although this was superior to the guttie it took several years before it was universally adopted.

Professional golfers now use a golf ball of 1.62 inches diameter in Britain (1.68 inches in the USA and Canada) and 1.62 oz maximum mass. This ball is made by winding elastic thread under tension around a central core which is then covered with a dimpled plastic coating.

Golf is a very popular game. It is likely to become even more popular as more farmland is converted to leisure pursuits. Pleasure-seeking mathematicians have much to ponder as they wait their turn at the tee.

### 13.2   AERODYNAMICS OF THE GOLF BALL

In Chapter 5 we described how the law of drag was related to the Reynolds number of the flow. For a smooth sphere we know that for $500 < \mathrm{Re} < 10^5$ the drag is proportional to the square of the speed, and the coefficient of drag $C_D$, has a constant value, about 0.45. At Reynolds numbers higher than about $2 \times 10^5$ the drag coefficient reduces to about 0.18, so that it is possible for an object to experience reduced drag even as its speed increases. The drop in $C_D$ was explained by the phenomenon of turbulence: when the Reynolds number exceeds a certain critical value, the smooth flow in the boundary layer becomes turbulent, the effect of which is to delay the boundary layer separation and reduce the drag. If the body surface is not particularly smooth the critical value of Re is reduced and so a rough body can experience reduced drag at Reynolds numbers for which the smooth counterpart would not.

A golf ball given a typical driving velocity of $70 \text{ ms}^{-1}$ has a value of Re of some 184,000 which is less than the critical value for a smooth sphere. However, a golf ball is not a smooth sphere but is heavily dimpled, the effect of which is to make the boundary layer flow turbulent, and so reduce the drag. A consequence of this is that a dimpled golf ball travels much further than the early smooth versions.

One important way in which the flight of a golf ball differs from the flight of a shot is that a golfer imparts spin to the ball which, as you will recall from Chapter 5, can cause lift to be generated which in turn can sustain the ball's flight. Bottom spin is imparted if the ball is hit with a sloping foot or lofted club as illustrated in Fig. 13.1.

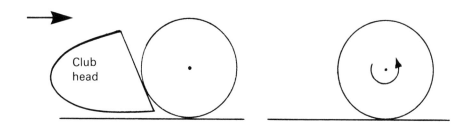

Fig. 13.1 — Backspin imparted by a lofted club.

The effect of this spin is to drag air around with the ball and so cause the boundary layer to be displaced. The lack of symmetry of the air flow causes a net force, or lift (sometimes called the Magnus effect). It opposes the effect of gravity and tends to keep the ball in the air for a longer time, thus extending the distance travelled before it hits the ground. Again, the dimples help reinforce this effect. In this chapter we present three different ways in which mathematicians have tried to model the golf ball drive.

## 13.3  MODEL 1

We will assume the drag and lift forces depend upon the velocity, according to the formulae $D = k_1 V^\alpha$ and $L = k_2 V^\beta$ where $k_1, k_2, \alpha$ and $\beta$ are constants which must be determined experimentally. There can be substantial variation in these constants, depending upon the presence or lack of spin, the spin rate, etc. Davies (1949) obtained values for the constants under a variety of conditions. For a dimpled ball spinning at 4000 rpm he obtained $L = 5.834 \times 10^{-3} V$ and $D = 3.589 \times 10^{-4} V^2$ and we will use these data for our model.

## 13.4  THE EQUATIONS OF MOTION

The forces acting on the golf ball are shown in Fig. 13.2.

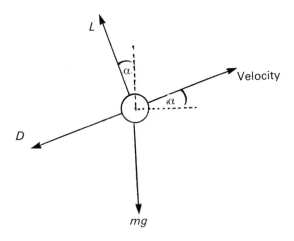

Fig. 13.2 — Forces acting on the golf ball.

Applying Newton's second law of motion horizontally and vertically gives

$$\frac{d^2x}{dt^2} = \frac{du}{dt} = -\frac{D\cos\alpha}{m} - \frac{L\sin\alpha}{m}$$

$$\frac{d^2y}{dt^2} = \frac{dw}{dt} = \frac{L\cos\alpha}{m} - \frac{D\sin\alpha}{m} - g$$

where $\tan\alpha = w/u$, and $L$ and $D$ are given by the formulae in the last section, and $m$, the mass of the golf ball, is 0.0459 kg.

In order that these equations can be solved numerically, we write them as four first-order equations and use the fourth-order Runge–Kutta method (see Computer Program 7 in Appendix III).

### 13.5   RESULTS

The time, $x$ coordinate and height of the golf ball are shown in Table 13.1 for initial

**Table 13.1**

| $t$ | $x$ | $y$ |
|---|---|---|
| 1 | 47.7 | 8.8 |
| 2 | 81.7 | 11.0 |
| 3 | 108.7 | 7.3 |
| 4 | 131.4 | − 1.7 |

data $\theta = 12°$, $V_0 = 60$ ms$^{-1}$. The range is 128.2 m and the time of flight is 3.8 s. The trajectory is plotted in Fig. 13.3.

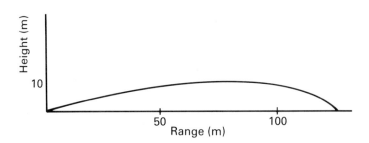

Fig. 13.3 — Trajectory of the golf ball.

Note that the initial flight path is roughly a straight line because the lift due to the bottom spin balances the weight. This behaviour is characteristic of actual golf ball trajectories.

### 13.6   MODEL 2: AN ANALYTICAL SOLUTION

As in section 13.3 we assume that $D = k_1 V^2$ and $L = k_2 V$, but instead of writing down the equations of motion in the usual way we will use intrinsic coordinates as described in section 4.5.
       The intrinsic equations of motion are

$$\frac{dV}{dt} = -\frac{k_1}{m} V^2 - g\sin\alpha$$

$$V\frac{d\alpha}{dt} = -g\cos\alpha + \frac{k_2 V}{m}$$

In the case of a normal golf drive the angle of projection is small and it is possible to make some analytical progress by assuming $\cos\alpha \approx 1$ and $\sin\alpha \approx 0$ (Tait, 1893). The equations of motion become

$$\frac{dV}{dt} = -\frac{k_1}{m} V^2 \tag{13.1}$$

$$V\frac{d\alpha}{dt} = -g + \frac{k_2}{m} V \tag{13.2}$$

Equation (13.1) can be solved by the method of separation of variables:

$$\int\frac{dV}{V^2} = \int\frac{-k_1}{m} dt$$

$$\frac{1}{V} - \frac{1}{V_0} = \frac{k_1}{m} t \tag{13.3}$$

where $V_0$ is the initial velocity at $t = 0$. Substituting for $V$ in (13.2) gives

$$\frac{d\alpha}{dt} = \frac{k_2}{m} - g\frac{k_1}{m} t - \frac{g}{V_0}$$

which gives, after integrating

$$\alpha = \theta + \left(\frac{k_2}{m} - \frac{g}{V_0}\right) t - \frac{gk_1}{2m} t^2 \tag{13.4}$$

where $\theta$ is the initial angle of projection.

Now if the golf ball moves a distance $\delta s$, its approximate increase in height is $\alpha \delta s$ and its approximate change in $x$ coordinate is $\delta s$, if $\alpha$ is small. Therefore

$$x = \int_0^s ds$$

$$= \int_0^t \frac{ds}{dt} dt$$

$$= \int_0^t V dt$$

and

$$y = \int_0^s \alpha \, ds$$

$$= \int_0^t \alpha \frac{ds}{dt} dt$$

$$= \int_0^t \alpha V \, dt$$

Therefore, from (13.3) and (13.4) we have

$$x = \int_0^t \frac{V_0 m}{m + k_1 V_0 t} dt$$

$$y = \int_0^t \left[ \frac{\theta V_0 m}{m + k_1 V_0 t} + \frac{(k_2 V_0 - gm)t}{m + k_1 V_0 t} - \frac{g k_1 t^2 V_0}{2(m + k_1 V_0 t)} \right] dt$$

Integration of these equations gives

$$x = \frac{m}{k_1} \log_e \left( \frac{m + k_1 V_0 t}{m} \right) \tag{13.5}$$

$$y = \left(\frac{\theta m}{k_1} - \frac{m^2}{k_1^2 V_0^2}\left(\frac{V_0 k_2}{m} - g\right) - \frac{m^2 g}{2k_1^2 V_0^2}\right) \log_e \left(\frac{m + k_1 V_0 t}{m}\right)$$

$$+ \frac{m\left(2V_0\frac{k_2}{m} - g\right)t}{2k_1 V_0} - \frac{gt^2}{4} \tag{13.6}$$

Using (13.5) and (13.6) we can calculate the height and horizontal displacement of the golf ball throughout its flight and so plot the trajectory. The results are shown in Table 13.2 using the data of section 13.3. The time of flight is 3.9 s and the range is

**Table 13.2**

| $t$ | $x$ | $y$ |
|---|---|---|
| 1 | 49.2 | 8.9 |
| 2 | 84.6 | 11.3 |
| 3 | 112.4 | 8.0 |
| 4 | 135.1 | − 0.8 |
| 5 | 154.5 | − 14.9 |

133 m. The data in Tables 13.1 and 13.2 should be compared.

## 13.7  MODEL 3

The coefficient of drag for a smooth sphere experiences a sudden fall from 0.45 to 0.18 when the Reynolds number, Re, reaches a certain critical value. Experimental work shows that if the surface of the sphere is rough, the critical value is reduced and the fall in $C_D$ takes place over a wider speed range. Williams (1959) has shown that the relationship is of the form

$$C_D = 14/V \tag{13.7}$$

for velocities ranging from 55 to 84 ms$^{-1}$ which represent the practical range of speeds in golf ball flight. As a consequence the total drag increases only linearly with $V$ instead of as the velocity squared.

Williams was interested in the question of whether a professional who achieves a longer drive than an ordinary golfer by projecting the ball at a higher speed, is taking advantage of this drop in $C_D$ and is thereby obtaining 'something for nothing'.

At 58 ms$^{-1}$ (a typical initial velocity for an average player), $C_D = 0.241$ whilst at 69 ms$^{-1}$ (typical of a long hitter), $C_D = 0.203$ from (13.7). He shows that if both players were to have a drag coefficient fixed at the higher value, the distance travelled by the faster ball during the time its velocity drops from 69 to 58 ms $^{-1}$ is 39 m. The calculated extra carry is about 48 m, so that some 9 m is due to the drop in $C_D$. Therefore although the long hitter does achieve something for nothing, the majority of the gain is due to honest effort.

## INVESTIGATION

Re-examine this model numerically: given $m = 0.046$ kg, and the radius $r = 0.02$ m, modify the computer program so that $C_D = 14/V$. With an angle of projection of 6° and the given projection speeds, find the ranges of the average and professional players. How much of the increase in range is due to the fall in $C_D$?

# 14

## High-altitude ballistics

### 14.1 INTRODUCTION

In all the previous models it has been assumed that the projectile does not rise very high above the surface of the Earth. In some applications, however, this assumption is not valid. Long-distance artillery is one example where the projectile may rise to a height of several kilometres. During the First World War it was discovered that if a shell were fired at an angle greater than the theoretical optimal angle for maximum range then an even greater range was obtained. If the angle of projection is high then the shell will reach high altitudes where the air resistance is small because of the reduced density of the air. Recall that the drag force, $D$, is given by

$$D = \tfrac{1}{2}C_D\rho A V^2$$

and so for any given shell moving at a velocity $V$ the drag is proportional to the air density. This fact enables the gunner to achieve large ranges.

In this chapter we shall devise a mathematical model of this phenomenon.

### 14.2 THE MODEL

We will make the following assumptions:

(i)   The shell is a steel sphere of radius 0.15 m, density 8000 kgm$^{-3}$, which is projected at 800 ms$^{-1}$.
(ii)  Although at such high Reynolds number, Re, the drag coefficient is a complicated function of Re and also the Mach number (defined as the speed of the projectile divided by the speed of sound in the same medium), for ease of understanding we take $C_D$ to be constant and equal to 0.4, so that the drag is proportional to the square of the speed.
(iii) The density of the atmosphere depends upon many factors such as pressure and

temperature (Wallace & Hobbs, 1977). In this model we will assume that the most significant variation of $\rho$ is with height above the surface of the Earth. It is true to say that $\rho$ also varies horizontally, but the vertical variation is much larger. Data obtained from various measuring devices, such as high-altitude balloons allowed to rise in the atmosphere, show that the variation of $\rho$ with height, $y$, is typically that illustrated in Fig. 14.1.

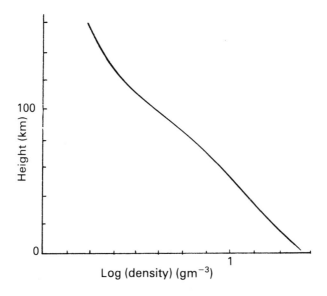

Fig. 14.1 — Variation of density with height in the atmosphere.

Since the height–log(density) curve is very nearly a straight line it is reasonable to assume a relationship of the form

$$y = -k\log_e \rho/\rho_0$$

where $k$ is a constant and $\rho_0$ is the density when $y = 0$, i.e. at sea-level. An approximate value of $k$ obtained from such a graph is 8500, so that we can take

$$\rho = \rho_0 \, e^{-y/8500}$$

The reader can verify that when $y$ is about 5 km the density is reduced to little more than half its value at sea-level which we take to be 1.22 kgm$^{-3}$.

In fact half the mass of the atmosphere lies below 5.5 km (less than 0.001 times the radius of the Earth), and about 99% of the mass lies within the lowest 30 km above sea-level.

(iv) The projectiles we will consider will rise to heights of some 5 km. We noted in Chapter 1 that $g$, the acceleration due to gravity, itself varies with height, but here we will neglect the effects of such change. Nothing essential is lost by this approximation since in 5 km, $g$ only alters by some 0.15%.

## 14.3  EQUATIONS OF MOTION

Recall from Chapter 5 that the drag force can be written

$$D = -\tfrac{1}{2}\rho(y)C_DAV^2\hat{\mathbf{V}}$$

$$= -\tfrac{1}{2}\rho(y)C_DAV\begin{pmatrix} u \\ w \end{pmatrix}$$

where $V = \sqrt{(u^2 + w^2)}$.

The equations of motion are then

$$\ddot{x} = -\frac{\rho(y)C_DAVu}{2m}$$

$$\ddot{y} = -\frac{\rho(y)C_DAVw}{2m} - g$$

where $m$ is the mass of the cannon ball. In this case $A$, the area of the projection of the sphere onto the free stream, is $\pi r^2$, which is equal to 0.070. With $C_D = 0.4$, $g = 9.8$ ms$^{-2}$ and $m = 113.1$ kg we have

$$\ddot{x} = -1.25 \times 10^{-4}\, \rho(y)(u^2 + w^2)^{1/2}u$$

$$\ddot{y} = -1.25 \times 10^{-4}\, \rho(y)\, (u^2 + w^2)^{1/2}w - 9.8$$

with $\rho(y)$ as described in section 14.2.

These two second-order differential equations must be solved numerically by expressing them as four first-order equations. Computer Program 8 in Appendix III does this when it is provided with a projection angle, $\theta$, and then outputs the range. The initial conditions are $u_0 = 800\cos\theta$, and $w_0 = 800\sin\theta$.

## 14.4  RESULTS

We first present in Table 14.1 some results of running the program with $\rho(y) = $ constant $= 1.22$ kgm$^{-3}$.

It is apparent from the table that the maximum range is about 11.038 km when $\theta$ is about 33°. When we allow the density to vary we obtain the results in Table 14.2.

Note that the maximum range is now increased to about 13.309 km and the corresponding angle of projection is about 43°.

It is now clear why aiming the cannon at an angle higher than what had been believed to give maximum range resulted in an increased range. The resistance data

**Table 14.1** — Range and angle of projection when $\rho$ = constant

| $\theta°$ | Range (km) |
|---|---|
| 28 | 10.946 |
| 29 | 10.982 |
| 30 | 11.009 |
| 31 | 11.027 |
| 32 | 11.037 |
| 33 | 11.038 |
| 34 | 11.031 |
| 35 | 11.016 |
| 36 | 10.993 |
| 37 | 10.962 |
| 38 | 10.924 |

$\rho(y) = 1.22 \text{ kgm}^{-3}$

**Table 14.2** — Range and angle of projection when $\rho = \rho_0 e^{-y/8500}$

$\rho(y) = 1.22 \, e^{-y/8500} \text{ kgm}^{-3}$

| $\theta°$ | Range (km) |
|---|---|
| 38 | 13.201 |
| 39 | 13.241 |
| 40 | 13.272 |
| 41 | 13.294 |
| 42 | 13.306 |
| 43 | 13.309 |
| 44 | 13.302 |
| 45 | 13.285 |
| 46 | 13.258 |
| 47 | 13.220 |
| 48 | 13.172 |

had been gathered for layers of air near to the Earth's surface whereas the shell in our example rises to a maximum height of 4.892 km where the density is reduced by some 44%.

**EXERCISE**

Modify the program in the appendix to take account of motion of the air, i.e. wind.
(Hint: see section 5.6.)

# Appendix I
# Vectors and differential equations

## I.1 VECTORS

Certain physical quantities are fully described by a single number, e.g. the mass of a stone, the speed of a car. Such quantities are called *scalars*. On the other hand, some quantities are not fully described until a direction is specified in addition to the number. For example, a velocity of $30\,\text{ms}^{-1}$ due east is different from a velocity of $30\,\text{ms}^{-1}$ due north. These quantities are called *vectors*. In our study of projectile motion it is necessary to distinguish between the two types.

Scalars are the simplest: the specification of a single number is all that is required. Vectors have a direction, and it is useful to consider a graphical representation. Thus the line segment AB of length 4 cm can represent a vector in the direction shown by the arrow on AB (Fig. A1). This vector is denoted by $\overrightarrow{AB}$. Note that $\overrightarrow{AB} \neq \overrightarrow{BA}$. An

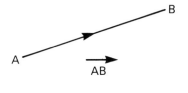

Fig. A1.

alternative notation is frequently used: we denote $\overrightarrow{AB}$ by **a**. The length of the line segment represents the magnitude or modulus of the vector, and we use the notation $|\overrightarrow{AB}|$, or $|\mathbf{a}|$ or sometimes $a$, to denote this.

The vector $-\mathbf{a}$ is a vector in the opposite direction to, but with the same magnitude as, **a**. Geometrically it will be $\overrightarrow{BA}$.

A unit vector is a vector whose length is equal to 1. If **a** has length 3, a unit vector in the direction of **a** is $\frac{1}{3}$**a**. We denote the unit vector in the direction of **a** by **â** (Fig. A2).

Fig. A2.

More generally

$$\hat{\mathbf{a}} = \frac{\mathbf{a}}{a}$$

Vector addition is defined using the **triangle law** (Fig. A3).

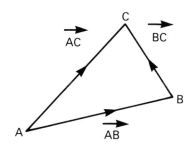

Fig. A3 — Vector addition.

We define the sum of $\overrightarrow{AB}$ and $\overrightarrow{BC}$ to be $\overrightarrow{AC}$.

Similarly if $\mathbf{r}_1$ and $\mathbf{r}_2$ are as shown, then the vector **b** represented by the third side of the triangle must be such that $\mathbf{r}_2 + \mathbf{b} = \mathbf{r}_1$ (Fig. A4). Thus $\mathbf{b} = \mathbf{r}_1 - \mathbf{r}_2$.

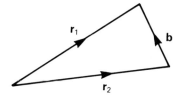

Fig. A4.

**Motion in the plane**

When we are dealing with two-dimensional projectile motion we can join the origin to the coordinates of any point $P(x,y)$ by a vector $\overrightarrow{O\,P}$ as shown (Fig. A5).

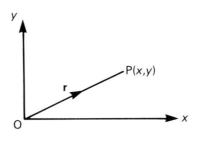

Fig. A5.

$\overrightarrow{O\,P}$ is called the position vector of P. Sometimes we will denote $\overrightarrow{O\,P}$ by **r**. The modulus of **r** is $|\mathbf{r}| = r$ and is the length of $\overrightarrow{O\,P}$.

If a projectile moves from P to Q then its position vector changes. If $\mathbf{r}_P$ and $\mathbf{r}_Q$ denote the two position vectors respectively, then we define the displacement vector of the projectile to be $\mathbf{r}_Q - \mathbf{r}_P$ as shown in Fig. A6.

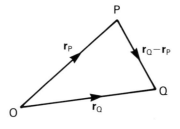

Fig. A6.

It is possible to express **r** in terms of the numbers $x$ and $y$. If we denote a unit vector along the $x$ axis by **i** and a unit vector along the $y$ axis by **j** then it is clear from the triangle law of addition that

$$\overrightarrow{O\,P} = \mathbf{r} = x\mathbf{i} + y\mathbf{j}$$

Then $r = \sqrt{x^2 + y^2}$ by Pythagoras' theorem.

An alternative notation which is sometimes useful is to write

$$\mathbf{r} = \begin{pmatrix} x \\ y \end{pmatrix}$$

This vector is called a column vector. Alternatively $\mathbf{r} = (xy)^{\mathrm{T}}$.

### Three-dimensional motion

The above notation generalises in a straightforward way to three-dimensional projectile motion. Taking Cartesian axes $x,y,z$ as shown in Fig. A7, and $\mathbf{i,j,k}$ to

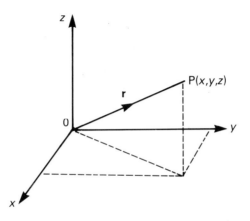

Fig. A7.

represent unit vectors in the direction of these axes, we can write the vector from O to any point $P(x,y,z)$ as

$$\mathbf{r} = x\mathbf{i} + y\mathbf{j} + z\mathbf{k}$$

Students should verify that

$$r = \sqrt{x^2 + y^2 + z^2}$$

### Velocity and acceleration

If a projectile moves along a curve in two dimensions its position vector will be dependent upon the time $t$. Suppose it is at point P at time $t$ and at point Q at a later time $t + \delta t$ (Fig. A8). Its average velocity in the time interval $t$ to $t + \delta t$ is its displacement vector divided by the time interval. Thus

$$\text{Average velocity} = \frac{\overrightarrow{PQ}}{\delta t} = \frac{\mathbf{r}(t + \delta t) - \mathbf{r}(t)}{\delta t}$$

The instantaneous velocity at time $t$ is then

$$\mathbf{v} = \lim_{\delta t \to 0} \frac{\mathbf{r}(t + \delta t) - \mathbf{r}(t)}{\delta t}$$

Now $\mathbf{r}(t) = x(t)\mathbf{i} + y(t)\mathbf{j}$ so $\mathbf{r}(t + \delta t) = x(t + \delta t)\mathbf{i} + y(t + \delta t)\mathbf{j}$. Therefore

$$\mathbf{v} = \lim_{\delta t \to 0} \frac{x(t + \delta t)\mathbf{i} + y(t + \delta t)\mathbf{j} - x(t)\mathbf{i} - y(t)\mathbf{j}}{\delta t}$$

$$= \frac{dx\mathbf{i}}{dt} + \frac{dy\mathbf{j}}{dt}$$

which we denote by $\dot{\mathbf{r}}$, since $\mathbf{i}$ and $\mathbf{j}$ are fixed vectors independent of $t$. So the velocity vector $\mathbf{v} = \dot{\mathbf{r}}$ is the derivative of the position vector with respect to time. This generalises in an obvious way to three-dimensional motion. The magnitude of the velocity vector, $\mathbf{v}$, is the speed of the projectile. We define the acceleration in a similar way:

$$\mathbf{a} = \frac{d\mathbf{v}}{dt} = \frac{d^2\mathbf{r}}{dt^2} = \ddot{\mathbf{r}}$$

**Integration**
If a vector depends upon the time $t$ it will be frequently necessary to integrate with respect to time. Since $\mathbf{i}$ and $\mathbf{j}$ are constant vectors, there is no difficulty and the integral

$$\mathbf{I} = \int (f(t)\mathbf{i} + g(t)\mathbf{j}) dt$$

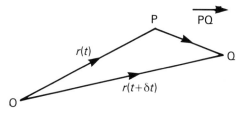

Fig. A8.

is simply evaluated as two scalar integrals:

$$\mathbf{I} = \int f(t)\mathrm{d}t\,\mathbf{i} + \int g(t)\mathrm{d}t\,\mathbf{j}$$

## I.2 DIFFERENTIAL EQUATIONS

The analytical techniques required for the solution of the differential equations encountered in this book are relatively simple. For completeness we state the main results here. Problems involving projectile motion generally reduce to the solution of the equation obtained by applying Newton's second law of motion:

$$\mathbf{F} = m\ddot{\mathbf{r}}$$

where $m$ is the mass of the projectile, $\mathbf{F}$ is the vector sum of all the forces acting on it and $\mathbf{r}$ is its position vector. The two dots denote that $\ddot{\mathbf{r}}$ is the second derivative of the position vector with respect to time, i.e. the acceleration. Because of the presence of the second derivative this is called a second-order equation.

This equation is equivalent to two (or three) scalar equations:

$$F_x = \frac{m\mathrm{d}^2 x}{\mathrm{d}t^2}, \; F_y = \frac{m\mathrm{d}^2 y}{\mathrm{d}t^2} \left( \text{and } F_z = \frac{m\mathrm{d}^2 z}{\mathrm{d}t^2} \text{ in 3 dimensions} \right)$$

where $\mathbf{F} = F_x\mathbf{i} + F_y\mathbf{j} + F_z\mathbf{k}$. It is therefore convenient to deal with the scalar equations separately. Sometimes an analytical solution is possible, but on occasions the equations are coupled (i.e. not independent) and we must resort to numerical techniques.

It is possible to express a second-order scalar differential equation as two first-order equations. Thus if we have

$$\frac{d^2x}{dt^2} = f(\dot{x}, x, t)$$

we can introduce the new variables $y_1$ and $y_2$ such that $y_1 = \dot{x}$ and $y_2 = x$, and then the equation becomes

$$\frac{dy_1}{dt} = f(y_1, y_2, t)$$

$$\frac{dy_2}{dt} = y_1$$

The problem then is to solve these two coupled first-order equations. Note that if $f(y_1, y_2, t)$ is independent of $y_2$ then

$$\frac{dy_1}{dt} = f(y_1, t), \qquad \frac{dy_2}{dt} = y_1$$

and the equations are uncoupled. The first can be solved and its solution used to find $y_2$. Such an example occurs when we study resisted projectile motion in Chapter 4.

**Variables separable**

Because many of the equations encountered can be reduced to first order we concentrate on these. Consider the equation

$$\frac{dx}{dt} = f(x, t)$$

If $f(x, t)$ can be written $f(x, t) = X(x)T(t)$, where $X(x)$ is a function of $x$ only and $T(t)$ is a function of $t$ only, the right-hand side is said to be *separable*.

Then

$$\int \frac{dx}{X(x)} = \int T(t)dt$$

We can now integrate both sides to obtain

$$\int \frac{dx}{X(x)} = \int T(t)dt + C$$

where $C$ is the constant of integration.

For example:

Solve $v\dfrac{dv}{dx} = x$

subject to the condition $v = v_0$ when $x = 0$.
This equation is separable. We obtain

$$\int v \, dv = \int x \, dx$$

Therefore

$$\frac{v^2}{2} = \frac{x^2}{2} + C$$

Applying the given condition we obtain $C = v_0^2/2$. Therefore

$$v^2 = x^2 + v_0^2$$

# Appendix II
## Numerical solution of differential equations

The complexity of some of the differential equations encountered in this book is such that no analytical treatment is possible and numerical methods must therefore be used. Satisfactory results can usually be obtained using the fourth-order Runge–Kutta method, and in this appendix we state its formulae. For a full theoretical treatment the reader is referred to a numerical analysis book. Consider the first-order differential equation

$$\frac{dx}{dt} = f(x,t) \qquad\qquad\qquad (II.1)$$

Given a starting point $t = t_0$ where the function $x$ has a value $x_0$ we wish to find $x(t)$ for $t > t_0$ that satisfies (II.1). A problem such as this is called an *initial-value problem*. In order to carry out a numerical solution the axis of the independent variable $t$ is divided into evenly spaced intervals of width $h$ whose endpoints are situated at $t_i = t_0 + ih \; i = 0,1,2. \ldots$. We denote the solution at $t_i$ by $x_i$. Note that we know that $x = x_0$ at $t = t_0$ already. We now calculate the function $x_1$ at $t_1$ from given formulae, the form of which depends upon the numerical method chosen. The fourth-order Runge–Kutta formulae are

$$x_1 = x_0 + (k_1 + 2k_2 + 2k_3 + k_4)/6$$

where

$$k_1 = hf(x_0,t_0)$$
$$k_2 = hf(x_0 + \tfrac{1}{2}k_1, t_0 + \tfrac{1}{2}h)$$
$$k_3 = hf(x_0 + \tfrac{1}{2}k_2, t_0 + \tfrac{1}{2}h)$$
$$k_4 = hf(x_0 + k_3, t_1)$$

Since $t_0$, $x_0$ and $f(x,t)$ are known, $x_1$ can be calculated. Knowing $x_1$ at $t_1$, the whole process can then be repeated to find $x_2$ at $t_2$, and so on.

This method can be extended in a natural way to deal with a system of $n$ first-order equations: suppose that

$$\frac{dx}{dt} = \mathbf{f}(\mathbf{x},t)$$

where $\mathbf{x} = (x_1 x_2 \ldots x_n)^T$ and $\mathbf{f} = (f_1 f_2 \ldots f_n)^T$, then given $\mathbf{x}_0$ at $t = t_0$ the vector $\mathbf{x}_1$ at $t_1$ is obtained from the formulae

$$\mathbf{x}_1 = \mathbf{x}_0 + (\mathbf{k}_1 + 2\mathbf{k}_2 + 2\mathbf{k}_3 + \mathbf{k}_4)/6$$

where

$$\mathbf{k}_1 = h\mathbf{f}(\mathbf{x}_0,t_0)$$
$$\mathbf{k}_2 = h\mathbf{f}(\mathbf{x}_0 + \tfrac{1}{2}\mathbf{k}_1, t_0 + \tfrac{1}{2}h)$$
$$\mathbf{k}_3 = h\mathbf{f}(\mathbf{x}_0 + \tfrac{1}{2}\mathbf{k}_2, t_0 + \tfrac{1}{2}h)$$
$$\mathbf{k}_4 = h\mathbf{f}(\mathbf{x}_0 + \tfrac{1}{2}\mathbf{k}_3, t_1)$$

$\mathbf{x}_1$ can thus be calculated. The process is then repeated to find $\mathbf{x}_2$, and so on. In this way the solution is obtained by 'marching forward' in time. To ensure a satisfactory solution the interval width $h$ must be sufficiently small. In the programs given in Appendix III it may be necessary to reduce $h$ for certain problems.

# Appendix III
# Computer programs

Computer programming is now beginning to achieve acceptance as a recognised component in mathematical education. The microcomputer is certainly widely used in many undergraduate courses in mathematics, at least as a visual aid. The visual display unit can act as an electronic blackboard, with the ability to display complex diagrams and numerical solutions to equations that are not amenable to analytic techniques.

The programs that are included in this appendix are all used to illustrate the text of the book in the way described above. Each program is independent of the others (although many perform similar tasks) and is self-explanatory. For the reasons given below, we have not included a floppy disk, so the reader will need to type the instructions into the microcomputer and store the program in the usual way.

It is a feature of these programs that they are interactive so the reader can readily understand the workings and thus make alterations to parameters or to the nature of the output.

The principle of easy interaction means that the programming language has to be one that is interpreted (like BASIC or LOGO) rather than a compiled language (like FORTRAN or PASCAL). The language needs to be one that is accessible to most readers with their own microcomputers. At the present time, most personal computer (PC) manufacturers supply a version of BASIC which is in the form of a ROM plugged inside the case.

We have decided to use BBC BASIC since this appears to be setting the standard for more recent versions of the language. However, the programs are written in a style which makes them as readable as possible and easily convertible to other versions of BASIC. To make the programs usable by a wider class of readers than just BBC micro owners, we use a restricted range of commands that can easily be converted into equivalents for other popular PCs such as AppleMac, the IBM PC, Sinclair machines, Apple II, and Research Machines 380/480Z.

For further information about BASIC, its applications to mathematics, and the relationship between various versions of BASIC, the reader is referred to Oldknow

(1984) and Oldknow (1987). Both are in the Ellis Horwood Series: Mathematics and its Applications.

## LIST OF PROGRAMS

No. 1    Plots trajectories for a projectile suffering no resistance, for a given initial speed and differing values of θ.

No. 2    Plots trajectories when $R \propto V$ for a given initial speed and different values of θ.

No. 3    Plots a typical trajectory when $R \propto V^2$.

No. 4a,b Newton–Raphson iteration for the solution of Baiera's problem with $R \propto V$ as described in section 7.3.

No. 5a,b Solution of Baiera's problem with $R \propto V^2$ as described in section 7.4.

No. 6    Calculates the position and pitch of a javelin under the conditions described in Chapter 10.

No. 7    Solves the equations of motion of a golf ball and calculates the range as described in section 13.4.

No. 8    Solves the equations of motion of a projectile moving in a medium where the density varies with the height, and calculates the range as described in Chapter 14.

*Note:* Because of the type of printer used in the listings, we draw the reader's attention to the following symbols:

   ç as in Tç2 means T∧2
   ~ as in Y(I)~H means Y(I)<H
   §% as in §%=131850 means @%=131850

## Program No. 1
Main Program (10–400)
              Sets up initial values and stores values of $x$ and $y$.

PROCPLOT (410–630)
              Plots trajectories for angles of projection 30°, 45°, 60°, 75° and 90°.

PROCBP (640–780)
              Plots the enveloping parabola.

L.

```
   1 REM PROGRAM NO.1
   2
  10 CLS
 150 DIMX(300),Y(300)
 160 DIM O(300),P(300)
 170 MODE5
 180 speed=15:h=2
 190 G=9.81
 200 F1$="V0*T-G*(Tç2)/2"
 210 F2$="U0*T"
 220 PRINT"+++TRAJECTORIES+++":PRINT"      NO RESISTANCE"
 230 PRINT:PRINT"V=15::::::h=2"
 240 PRINT:PRINT"YELLOW CURVE IS E.P."
 250 PROCplot
 260 PROCbp
 400 END
 401
 402
 403
 410 DEF PROCplot
 420 VDU5
 430 GCOL 0,3
 440 FOR X=0 TO 30
 450 MOVE 40*X+100,100:PRINT "*":NEXTX
 460 FOR Y=0TO2:MOVE100,40*Y+100:PRINT"*":NEXT Y
 470 GCOL0,1
 480 FOR Q=30 TO 90 STEP15
 490 A=RAD(Q)
 500 V0=speed*SIN(A):U0=speed*COS(A)
 510 T=0:I=0
 520 REPEAT
 530 I=I+1
 540 X(I)=EVAL(F2$)
 550 Y(I)=EVAL(F1$)
 560 T=T+.02
 570 UNTIL Y(I)~-h
 580 FOR R=1 TO I-1
 590 O(R)=X(R):P(R)=Y(R)
 600 MOVE 40*X(R)+100,40*Y(R)+200:PRINT".":NEXTR
 610 NEXT Q
 630 ENDPROC
 631
 632
 633
 640 DEF PROCbp
 650 a=(speedç2)/G
 660 X=0:I=0
 670 GCOL 0,2
 680 REPEAT
 690 I=I+1
 700 X(I)=X
 710 Y(I)=((aç2)-(Xç2))/(a*2)
 720 X=X+.5
 730 UNTIL Y(I)~-h
 740 FOR R=1 TO I-1
 750 MOVE 40*X(R)+120,40*Y(R)+200
 760 PRINT"."
 770 NEXTR
 780 ENDPROC
..
```

**Program No. 2**
Main Program (130–380)
Sets up and stores values of *x* and *y* when *k*=0.1.

PROCPLOT (410–630)
Plots trajectories for angles of projection 30°, 40°, ..., 90°.

L.

```
   1 REM PROGRAM NO.2
   2
   3
 130 CLS
 150 DIMX(300),Y(300)
 160 DIM O(300),P(300)
 170 MODE5
 180 speed=10:K=.1:h=2
 190 t=9.81/K:G=9.81
 290 F2$="U0*(1-EXP(-K*T))/K"
 300 F1$="(((t+V0)*(1-EXP(-K*T)))/K)-t*T"
 310 PRINT"+++TRAJECTORIES+++"
 320 PRINT"RESISTANCE...K=.1"
 330 PRINT:PRINT"V=10:::::::h=2"
 360 PROCplot
 380 END
 381
 382
 383
 410 DEF PROCplot
 420 VDU5
 430 GCOL 0,3
 440 FOR X=0 TO 30
 450 MOVE 40*X+100,100:PRINT "*":NEXTX
 460 FOR Y=0TO4:MOVE100,40*Y+100:PRINT"*":NEXT Y
 470 GCOL0,1
 480 FOR Q=30 TO 90 STEP 10
 490 A=RAD(Q)
 500 V0=speed*SIN(A):U0=speed*COS(A)
 510 T=0:I=0
 520 REPEAT
 530 I=I+1
 540 X(I)=EVAL(F2$)
 550 Y(I)=EVAL(F1$)
 560 T=T+.02
 570 UNTIL Y(I)~-h
 580 FOR R=1 TO I-1
 590 O(R)=X(R):P(R)=Y(R)
 600 MOVE 80*X(R)+100,80*Y(R)+300:PRINT".":NEXTR
 610 NEXT Q
 630 ENDPROC
 ..
```

**Program No. 3**
Main Program (5–80)
Calculates current values of *x*, *y*, *u* and *w* using the fourth-order
Runge–Kutta method.
(550–560)

Define functions necessary for equations of motion.

**PROCINITIAL (100–170)**

Sets up the angle of projection, *x, y, u, w* and the value of *k*.

**PROCKUTTA (190–375)**

Calculates the Runge–Kutta parameters as defined in Appendix II.

**PROCPRIN (470–490)**

Prints headings, etc.

**PROCPLOT (510–545)**

Plots trajectory.

```
L.
    1 REM PROGRAM NO.3
    5 MODE4
    6 CLS
   20 PROCINITIAL
   30 REPEAT
   35 PROCPLOT
   40 PROCKUTTA(X,Y,U,W,T)
   41 T=T+H
   42 X=X+(D1X+2*(D2X+D3X)+D4X)/6
   43 Y=Y+(D1Y+2*(D2Y+D3Y)+D4Y)/6
   44 U=U+(D1U+2*(D2U+D3U)+D4U)/6
   45 W=W+(D1W+2*(D2W+D3W)+D4W)/6
   50 UNTIL Y~0
   60 PROCPRIN
   80 END
   81
   82
   83
  100 DEFPROCINITIAL
  105 THETA =30
  110 THE=THETA*3.14159/180
  115 K=.04:S=60
  120 X=0
  130 Y=0
  135 T=0
  140 U=S*COS(THE)
  150 W=S*SIN(THE)
  160 H=0.05
  170 ENDPROC
  171
  172
  173
  190 DEFPROCKUTTA(X,Y,U,W,T)
  200 D1X=H*U:D1Y=H*W
  210 D1U=H*FNF1(U,W,Y)
  220 D1W=H*FNF2(U,W,Y)
  240 D2X=H*(U+D1U/2)
  250 D2Y=H*(W+D1W/2)
  260 D2U=H*FNF1(U+D1U/2,W+D1W/2,Y+D1Y/2)
  270 D2W=H*FNF2(U+D1U/2,W+D1W/2,Y+D1Y/2)
  290 D3X=H*(U+D2U/2)
  300 D3Y=H*(W+D2W/2)
  310 D3U=H*FNF1(U+D2U/2,W+D2W/2,Y+D2Y/2)
```

```
320 D3W=H*FNF2(U+D2U/2,W+D2W/2,Y+D2Y/2)
340 D4X=H*(U+D3U)
350 D4Y=H*(W+D3W)
360 D4U=H*FNF1(U+D3U,W+D3W,Y+D3Y)
370 D4W=H*FNF2(U+D3U,W+D3W,Y+D3Y)
375 ENDPROC
376
377
378
470 DEFPROCPRIN
471 PRINT "RESISTED MOTION"
472 PRINT "R PROPORTIONAL TO V SQUARED":PRINT
480 PRINT "THETA =";THETA
481 PRINT "INITIAL SPEED = ";S
482 PRINT "K = ";K
490 ENDPROC
491
493
510 DEFPROCPLOT
520 VDU5:MOVE X*25,Y*25:PRINT "."
530 ENDPROC
540 VDU5:MOVE X*25,Y*25:PRINT "."
545 ENDPROC
546
547
548
550 DEFFNF1(U,W,Y)=-K*SQR(U*U+W*W)*U
560 DEFFNF2(U,W,Y)=-K*SQR(U*U+W*W)*W-9.8
..
```

## Program No. 4a
Main Program (10–70)

        Sets initial values of angle ($T$) and speed ($V$).
        (80–130)
        Define functions required to invert the Jacobian matrix

PROCNEW (140–210)

        Uses Newton–Raphson iteration to solve (7.3) and (7.4).

## Program No. 4b
Main Program (10–50)

        Sets initial values of angle ($T$) and estimate of range ($X$).
        (80–130)
        Define functions required to invert the Jacobian matrix.

PROCNEW (140–210)

        Uses Newton–Raphson iteration to solve (7.5) and (7.6).

## Program No. 5a
Main Program (10–190)

        Draws the inclined plane and marks the point A. Inputs starting speed

L.

```
  1 REM PROGRAM NO.4A
  2
 10 CLS: T=40: PRINT "INITIAL ESTIMATE OF ANGLE = ";T
 20 T=T*PI/180
 30 E=0.0001
 40 S%=131850
 50 V=24:PRINT "INITIAL ESTIMATE OF SPEED = ";V
 60 PRINT "PRESS ANY KEY TO CONTINUE";A$=GET$:PROCNEW
 70 END
 71
 72
 73
 80 DEFFNJ1(T,V)=59.963/(COS(T)*COS(T))+5996.3*SIN(T)*(1/(V*COS(T))-1/(COS(T)*V-5.9963))/COS(T)
 90 DEFFNJ2(T,V)=5996.3*(-1/(V*COS(T))+1/(V*COS(T)-5.9963))/V
100 DEFFNJ3(T,V)=5.9963*SIN(T)/(COS(T)*COS(T))+599.63/(COS(T)*COS(T))
110 DEFFNJ4(T,V)=5.9963/COS(T)-2*V
120 DEFFNF1(T,V)=59.963*TAN(T)+5996.3/(V*COS(T))+1000*LN(1-5.9963/(V*COS(T)))-0.59397
130 DEFFNF2(T,V)=5.9963*V/COS(T)+599.63*TAN(T)-V*V
131
132
133
140 DEFPROCNEW
150 DEL=FNJ1(T,V)*FNJ4(T,V)-FNJ2(T,V)*FNJ3(T,V)
160 TN=T-(FNJ4(T,V)*FNF1(T,V)-FNJ2(T,V)*FNF2(T,V))/DEL
170 VN=V-(FNJ1(T,V)*FNF2(T,V)-FNJ3(T,V)*FNF1(T,V))/DEL
180 IF ABS(VN-V)~E AND ABS(TN-T)~E THEN PRINT T*180/PI,V:ENDPROC
190 V=VN:T=TN
200 PRINT T*180/PI,V
210 PROCNEW
```

:

L.

```
1   REM PROGRAM NO.4B
2
10  CLS:T=40:PRINT "INITIAL ESTIMATE OF ANGLE = ";T
20  T=T*PI/180
30  E=0.0001
40  S%=131850
50  X=60:PRINT "INITIAL ESTIMATE OF RANGE = ";X
60  PRINT "PRESS ANY KEY TO CONTINUE":A$=GET$:PROCNEW
70  END
71
72
73
80  DEFFNJ1(T,X)=X/(COS(T)*COS(T))+3.6215*X*TAN(T)/COS(T)-(3.6215*X*TAN(T)/COS(T))/(1-0.0036215*X/COS(T))
90  DEFFNJ2(T,X)=TAN(T)+3.6215/COS(T)-3.6215/(COS(T))*(1-0.0036215*X/COS(T)))
100 DEFFNJ3(T,V)=V*2.7613*TAN(T)/COS(T)+10*V/(COS(T)*COS(T))
110 DEFFNJ4(T,X)=2.7613/COS(T)+10*TAN(T)
120 DEFFNF1(T,X)=X*TAN(T)+3.6215*X/COS(T)+1000*LN(1-0.0036215*X/COS(T))+1.5
130 DEFFNF2(T,V)=V*2.7613/COS(T)+10*V*TAN(T)-762.478
131
132
133
140 DEFPROCNEW
150 DEL=FNJ1(T,X)*FNJ4(T,X)-FNJ2(T,X)*FNJ3(T,X)
160 TN=T-(FNJ4(T,X)*FNF1(T,X)-FNJ2(T,X)*FNF2(T,X))/DEL
170 XN=X-(FNJ1(T,X)*FNF2(T,X)-FNJ3(T,X)*FNF1(T,X))/DEL
180 IF ABS(XN-X)^E AND ABS(TN-T)^E THEN PRINT T*180/PI,X:ENDPROC
190 X=XN:T=TN
200 PRINT T*180/PI,X
210 PROCNEW
    :
```

(*S*). Sends control to various procedures in order to plot several trajectories for $\theta = 30°, 35°, \ldots, 70°$.
(185–186)
Define functions which make up the equations of motion.

PROCINITIAL (200–280)

Sets up initial values of *x, y, u, w* and *t*, and defines time increment for the Runge–Kutta method (270)

PROCKUTTA (290–450)

Calculates the Runge–Kutta parameters as defined in Appendix II.

PROCQ (480–500)

Tests to see if the trajectory passes through the required point on the inclined plane.

PROCEND1 (510–530)

Stops program with displayed message if the speed is too large.

PROCPLOT (540–560)

Plots trajectory.

PROCEND2 (570–590)

Warns user that either a higher speed is required or a reduced step is needed in line 50.

*Note 1*. When the user's trajectory intercepts the plane near the required point, the step in line 50 should be modified.
*Note 2*. This program will take a long time to run.

```
L.
   1 REM PROGRAM NO.5A
   2
  10 MODE0
  30 MOVE 0,0:DRAW 1199,42:MOVE 1205,50:VDU5:PRINT "A":VDU4
  31 MOVE 0,0:DRAW 1199,0
  40 INPUT "INPUT INITIAL SPEED",S:PRINT "TRAJECTORIES ARE NOW BEING PLOTTED"
  50 FOR THETA=30 TO 70 STEP 40
  60 PROCINITIAL
  70 REPEAT
  80 PROCKUTTA(X,Y,U,W,T)
  90 T=T+H
 100 X=X+(D1X+2*(D2X+D3X)+D4X)/6
 110 Y=Y+(D1Y+2*(D2Y+D3Y)+D4Y)/6
 120 U=U+(D1U+2*(D2U+D3U)+D4U)/6
 130 W=W+(D1W+2*(D2W+D3W)+D4W)/6
 140 PROCQ
 150 PROCPLOT
 160 UNTIL Y~=-1.5AND W~0
 170 NEXT
 180 PROCEND2
```

```
185 DEFFNF1(U,W,Y)=-0.005*SQR(U*U+W*W)*U
186 DEFFNF2(U,W,Y)=-0.005*SQR(U*U+W*W)*W-9.8
190 END
191
192
193
200 DEFPROCINITIAL
210 THE=THETA*3.14159/180
220 X=0
230 Y=0
240 T=0
250 U=S*COS(THE)
260 W=S*SIN(THE)
270 H=.01
280 ENDPROC
281
282
283
290 DEFPROCKUTTA(X,Y,U,W,T)
300 D1X=H*U:D1Y=H*W
310 D1U=H*FNF1(U,W,Y)
320 D1W=H*FNF2(U,W,Y)
330 D2X=H*(U+D1U/2)
340 D2Y=H*(W+D1W/2)
350 D2U=H*FNF1(U+D1U/2,W+D1W/2,Y+D1Y/2)
360 D2W=H*FNF2(U+D1U/2,W+D1W/2,Y+D1Y/2)
370 D3X=H*(U+D2U/2)
380 D3Y=H*(W+D2W/2)
390 D3U=H*FNF1(U+D2U/2,W+D2W/2,Y+D2Y/2)
400 D3W=H*FNF2(U+D2U/2,W+D2W/2,Y+D2Y/2)
410 D4X=H*(U+D3U)
420 D4Y=H*(W+D3W)
430 D4U=H*FNF1(U+D3U,W+D3W,Y+D3Y)
440 D4W=H*FNF2(U+D3U,W+D3W,Y+D3Y)
450 ENDPROC
451
453
480 DEFPROCQ
490 IF X"=59.963 AND Y"=0.594 PROCEND1
500 ENDPROC
501
502
503
510 DEFPROCEND1
520 PRINT "WITH A SPEED  ";S;"  THE REQUIRED POINT CAN CERTAINLY BE REACHED."
521 PRINT "THIS MAY NOT BE THE MINIMUM SPEED. RE-RUN WITH A LOWER ONE"
530 END:ENDPROC
531
532
533
540 DEFPROCPLOT
550 PLOT 69,X*20,Y*20+21.5
560 ENDPROC
570 DEFPROCEND2
580 PRINT "RE-RUN WITH A HIGHER SPEED OR A REDUCED STEP IN LINE 50"
590 ENDPROC
591
592
593
```

**Program No. 5b**

Main Program (5–80)

> Defines a loop for $\theta = 41°$ to $48°$ and calculates the range for each angle.
> Prints out intermediate results.
> (510–520)
> Define functions which make up the equations of motion.

PROCINITIAL (100–170)

> Sets initial values of $x$, $y$, $u$ and $w$.

PROCKUTTA (190–375)

> Calculates the Runge–Kutta parameters as defined in Appendix II.

L.

```
  1 REM PROGRAM NO.5B
  2
  5 S%=131850
  8 S=27.63
  9 CLS:PRINT"INTEGRATIONS IN PROGRESS":PRINT:PRINT:PRINT
 10 FOR THETA=41 TO 48   STEP .5
 20 PROCINITIAL
 30 REPEAT
 40 PROCKUTTA(X,Y,U,W,T)
 41 T=T+H
 42 X=X+(D1X+2*(D2X+D3X)+D4X)/6
 43 Y=Y+(D1Y+2*(D2Y+D3Y)+D4Y)/6
 44 U=U+(D1U+2*(D2U+D3U)+D4U)/6
 45 W=W+(D1W+2*(D2W+D3W)+D4W)/6
 50 UNTIL Y~=1.5AND W~0
 55 H=-H:PROCKUTTA(X,Y,U,W,T)
 56 X1=X+(D1X+2*(D2X+D3X)+D4X)/6
 57 Y1=Y+(D1Y+2*(D2Y+D3Y)+D4Y)/6
 58 RA=(X1*Y-X*Y1)/(Y-Y1)
 59 PRINT "THETA= ";THETA,"RANGE=  ";RA
 65 NEXT
 80 END
 81
 82
 83
100 DEFPROCINITIAL
110 THE=THETA*3.14159/180
120 X=0
130 Y=0
135 T=0
140 U=S*COS(THE)
150 W=S*SIN(THE)
160 H=.05
170 ENDPROC
171
172
173
190 DEFPROCKUTTA(X,Y,U,W,T)
200 D1X=H*U:D1Y=H*W
210 D1U=H*FNF1(U,W,Y)
220 D1W=H*FNF2(U,W,Y)
240 D2X=H*(U+D1U/2)
250 D2Y=H*(W+D1W/2)
260 D2U=H*FNF1(U+D1U/2,W+D1W/2,Y+D1Y/2)
```

```
270 D2W=H*FNF2(U+D1U/2,W+D1W/2,Y+D1Y/2)
290 D3X=H*(U+D2U/2)
300 D3Y=H*(W+D2W/2)
310 D3U=H*FNF1(U+D2U/2,W+D2W/2,Y+D2Y/2)
320 D3W=H*FNF2(U+D2U/2,W+D2W/2,Y+D2Y/2)
340 D4X=H*(U+D3U)
350 D4Y=H*(W+D3W)
360 D4U=H*FNF1(U+D3U,W+D3W,Y+D3Y)
370 D4W=H*FNF2(U+D3U,W+D3W,Y+D3Y)
375 ENDPROC
376
377
378
510 DEFFNF1(U,W,Y)=-0.005*SQR(U*U+W*W)*U
520 DEFFNF2(U,W,Y)=-0.005*SQR(U*U+W*W)*W-9.8
```

## Program No. 6
Main program (10–120)

> Defines a matrix to hold current values of Runge–Kutta parameters. Sends control to various procedures in order to perform the Runge–Kutta method.

PROCIN (160–230)

> Sets all initial values and physical constants for the javelin.

PROCNEW(i) (270–360)

PROCK (480–530)

> Performs the Runge–Kutta integrations.

PROCLD (400–440)

> Calculates velocity, angle of attack, lift and drag.

PROCMARCH (570–650)

> Updates values of all variables.

PROCTERMIN (690–760)

> Concludes the flight and determines whether or not the landing is successful. Calculates the range.

PROCPICT (800–850)

> Plots the trajectory.

### *List of variables*

| | |
|---|---|
| Y1: | Horizontal component of velocity |
| Y2: | Vertical component of velocity |
| Y3: | Angular velocity |
| Y4: | Pitch |
| Y5: | $x$ value |
| Y6: | $y$ value |

H:       time step
ALPHA:  angle of attack
PHI:     $\tan^{-1}$ (Y2/Y1)

*Note:* only lines 170–190 need be altered for the other examples.

L.

```
  1 REM PROGRAM NO.6
  2
 10 MODE0:FLAG=1:VDU28,1,6,72,1:DIM K(4,6):T=0:G=9.81
 20 PROCIN
 30 REPEAT
 40 PROCNEW0:PROCLD:I=1:PROCK
 50 PROCNEW1:PROCLD:I=2:PROCK
 60 PROCNEW1:PROCLD:I=3:PROCK
 70 PROCNEW2:PROCLD:I=4:PROCK
 80 PROCMARCH
 90 IF FLAG=1 PROCPICT
100 FLAG=FLAG+1
110 UNTIL Y6+LE*SIN(Y4)~=0 :PROCTERMIN
120 END
130
140
150
160 DEFPROCIN
170 H=0.05:Y6=2:Y5=0
180 Y4=30*PI/180:Y3=0*PI/180:
190 V=30:PHI=30*PI/180
200 Y1=V*COS(PHI):Y2=V*SIN(PHI):ALPHA=Y4-PHI
210 LE=1.057:MI=0.42:d=.255:M=0.80625
220 CLS
230 ENDPROC
240
250
260
270 DEFPROCNEW0:Y10=Y1:Y20=Y2:Y30=Y3:Y40=Y4:Y50=Y5:Y60=Y6:ENDPROC
280
290 DEFPROCNEW1:Y10=Y1+K(I,1)*.5:Y20=Y2+K(I,2)*.5:Y30=Y3+K(I,3)*.5
300 Y40=Y4+K(I,4)*.5:Y50=Y5+K(I,5)*.5:Y60=Y6+K(I,6)*.5
310 ENDPROC
320
330
340 DEFPROCNEW2:Y10=Y1+K(3,1):Y20=Y2+K(3,2):Y30=Y3+K(3,3)
350 Y40=Y4+K(3,4):Y50=Y5+K(3,5):Y60=Y6+K(3,6)
360 ENDPROC
370
380
390
400 DEFPROCLD
410 V=SQR(Y10ç2+Y20ç2):PHI=ATN(Y20/Y10):ALPHA=Y40-PHI
420 IF ALPHA=0 THEN L=0 ELSE L=SGN(ALPHA)*5.61E-5*((ABS(ALPHA)*180/PI)ç1.34)*V*V
430 D=2.4E-4*EXP(5.157*ABS(ALPHA))*V*V
440 ENDPROC
450
460
470
480 DEFPROCK
490 K(I,1)=-H*(L*SIN(PHI)+D*COS(PHI))/M
500 K(I,2)=H*(L*COS(PHI)-D*SIN(PHI))/M-G*H
```

```
510 K(I,3)=-H*(L*COS(ALPHA)+D*SIN(ALPHA))*d/MI
520 K(I,4)=Y30*H:K(I,5)=Y1O*H:K(I,6)=Y2O*H
530 ENDPROC
540
570 DEFPROCMARCH
580 Y1=Y1+(K(1,1)+2*(K(2,1)+K(3,1))+K(4,1))/6
590 Y2=Y2+(K(1,2)+2*(K(2,2)+K(3,2))+K(4,2))/6
600 Y3=Y3+(K(1,3)+2*(K(2,3)+K(3,3))+K(4,3))/6
610 Y4=Y4+(K(1,4)+2*(K(2,4)+K(3,4))+K(4,4))/6
620 Y5=Y5+(K(1,5)+2*(K(2,5)+K(3,5))+K(4,5))/6
630 Y6=Y6+(K(1,6)+2*(K(2,6)+K(3,6))+K(4,6))/6
640 T=T+H
650 ENDPROC
660
670
680
690 DEFPROCTERMIN
700 IF Y4¨=0 PRINT "FOUL LANDING"
710 IF Y4¯0 PRINT "JAVELIN LANDED POINT FIRST"
720 X2=(Y6+LE*SIN(Y4))/TAN(PHI)
730 §%=131850
740 PRINT "RANGE=";Y5+LE*COS(Y4)-X2
750 PRINT "TIME OF FLIGHT=";T
760 ENDPROC
770
780
790
800 DEFPROCPICT
810 FLAG=-6
820 PLOT 69,Y5*13,Y6*13
830 PLOT 1,30*COS(Y4),30*SIN(Y4)
840 PLOT 1,-60*COS(Y4),-60*SIN(Y4)
850 ENDPROC
```

## Program No. 7
Main Program (6–80)

Calculates current values of *x*, *y*, *u* and *w* using the Runge–Kutta method.

(55–58)

Uses linear interpolation to calculate the range.

(510–520)

Define functions which make up the equations of motion.

PROCINITIAL (100–170)

Sets initial values of angle of projection, *x*, *y*, *u* and *w*.

*PROCKUTTA (190–375)*

Calculates Runge–Kutta parameters as defined in Appendix II.

PROCPRIN (470–490)

Prints the angle of projection and the range.

L.
```
  1 REM PROGRAM NO.7
  2
  6 PRINT"CALCULATIONS IN PROGRESS"
 20 PROCINITIAL
 30 REPEAT
 40 PROCKUTTA(X,Y,U,W,T)
 41 T=T+H
 42 X=X+(D1X+2*(D2X+D3X)+D4X)/6
 43 Y=Y+(D1Y+2*(D2Y+D3Y)+D4Y)/6
 44 U=U+(D1U+2*(D2U+D3U)+D4U)/6
 45 W=W+(D1W+2*(D2W+D3W)+D4W)/6
 50 UNTIL Y~0
 55 H=-H:PROCKUTTA(X,Y,U,W,T)
 56 X1=X+(D1X+2*(D2X+D3X)+D4X)/6
 57 Y1=Y+(D1Y+2*(D2Y+D3Y)+D4Y)/6
 58 RA=(X1*Y-X*Y1)/(Y-Y1)
 60 PROCPRIN
 80 END
100 DEFPROCINITIAL
105 THETA=12
110 THE=THETA*3.14159/180
120 X=0
130 Y=0
135 T=0
140 U=60*COS(THE)
150 W=60*SIN(THE)
160 H=0.05
170 ENDPROC
190 DEFPROCKUTTA(X,Y,U,W,T)
200 D1X=H*U:D1Y=H*W
210 D1U=H*FNF1(U,W,Y)
220 D1W=H*FNF2(U,W,Y)
240 D2X=H*(U+D1U/2)
250 D2Y=H*(W+D1W/2)
260 D2U=H*FNF1(U+D1U/2,W+D1W/2,Y+D1Y/2)
270 D2W=H*FNF2(U+D1U/2,W+D1W/2,Y+D1Y/2)
290 D3X=H*(U+D2U/2)
300 D3Y=H*(W+D2W/2)
310 D3U=H*FNF1(U+D2U/2,W+D2W/2,Y+D2Y/2)
320 D3W=H*FNF2(U+D2U/2,W+D2W/2,Y+D2Y/2)
340 D4X=H*(U+D3U)
350 D4Y=H*(W+D3W)
360 D4U=H*FNF1(U+D3U,W+D3W,Y+D3Y)
370 D4W=H*FNF2(U+D3U,W+D3W,Y+D3Y)
375 ENDPROC
470 DEFPROCPRIN
480 PRINT "THETA = ";THETA;"    RANGE = ";RA
490 ENDPROC
510 DEFFNF1(U,W,Y)=-0.127*W-7.82E-3*SQR(U*U+W*W)*U
520 DEFFNF2(U,W,Y)=0.127*U-7.82E-3*SQR(U*U+W*W)*W-9.8
```
..

## Program No. 8
Main Program (20–80)

> Calculates current values of $x$, $y$, $u$ and $w$ using the Runge–Kutta method.
> (510–520)
> Define functions which make up the equations of motion.

PROCINITIAL (100–170)
> Inputs initial angle of projection. Sets initial values of $x$, $y$, $u$ and $w$. Defines time increment.

PROCKUTTA (190–375)
> Calculates Runge–Kutta parameters as defined in Appendix II.

PROCPRIN (470–490)
> Prints angle of projection and the range.

*Note:* for the constant density case, the reader should change lines 510 and 520 as follows:

510 DEFFNF1(U,W,Y)=−1.25E–4*1.22*SQR(U*U+W*W)*U
520 DEFFNF2(U,W,Y)=−1.25E–4*1.22*SQR(U*U+W*W)*W–9.8

```
L.
   1 REM PROGRAM NO.8
   2
  20 PROCINITIAL
  21
  30 REPEAT
  31
  40 PROCKUTTA(X,Y,U,W,T)
  41 T=T+H
  42 X=X+(D1X+2*(D2X+D3X)+D4X)/6
  43 Y=Y+(D1Y+2*(D2Y+D3Y)+D4Y)/6
  44 U=U+(D1U+2*(D2U+D3U)+D4U)/6
  45 W=W+(D1W+2*(D2W+D3W)+D4W)/6
  50 UNTIL Y~=0 AND W~0
  55 H=--H:PROCKUTTA(X,Y,U,W,T)
  56 X1=X+(D1X+2*(D2X+D3X)+D4X)/6
  57 Y1=Y+(D1Y+2*(D2Y+D3Y)+D4Y)/6
  58 RA=(X1*Y-X*Y1)/(Y-Y1)
  60 PROCPRIN
  80 END
  81
  82
  83
 100 DEFPROCINITIAL
 101 CLS
 105 INPUT "INPUT ANGLE OF PROJECTION (0 - 90)" ,THETA
 110 THE=THETA*3.14159/180
 120 X=0
 130 Y=0
 135 T=0
 140 U=800*COS(THE)
 150 W=800*SIN(THE)
 160 H=.1
 170 ENDPROC
 171
 172
 173
 190 DEFPROCKUTTA(X,Y,U,W,T)
 200 D1X=H*U:D1Y=H*W
 210 D1U=H*FNF1(U,W,Y)
```

```
220 D1W=H*FNF2(U,W,Y)
240 D2X=H*(U+D1U/2)
250 D2Y=H*(W+D1W/2)
260 D2U=H*FNF1(U+D1U/2,W+D1W/2,Y+D1Y/2)
270 D2W=H*FNF2(U+D1U/2,W+D1W/2,Y+D1Y/2)
290 D3X=H*(U+D2U/2)
300 D3Y=H*(W+D2W/2)
310 D3U=H*FNF1(U+D2U/2,W+D2W/2,Y+D2Y/2)
320 D3W=H*FNF2(U+D2U/2,W+D2W/2,Y+D2Y/2)
340 D4X=H*(U+D3U)
350 D4Y=H*(W+D3W)
360 D4U=H*FNF1(U+D3U,W+D3W,Y+D3Y)
370 D4W=H*FNF2(U+D3U,W+D3W,Y+D3Y)
375 ENDPROC
376
377
378
470 DEFPROCPRIN
480 PRINT "THETA= ";THETA;"RANGE= ";RA
490 ENDPROC
491
492
493
510 DEFFNF1(U,W,Y)=-1.25E-4*EXP(-Y/8500)*1.22*SQR(U*U+W*W)*U
520 DEFFNF2(U,W,Y)=-1.25E-4*1.22*EXP(-Y/8500)*SQR(U*U+W*W)*W-9.8
```

# Appendix IV
## Selected annotated bibliography

As the foregoing chapters have shown, the field of 'projectiles' provides many examples of mathematical models for use in mechanics or mathematical modelling courses, and is a rich source of material suitable for undergraduate project work. There are numerous other published works which do likewise, and in this appendix we outline a selection of those which we hope may be of use to students and their tutors when looking for 'starting points'.

Baiera, J. C. (1976) The Physics of the Softball Throw. *The Physics Teacher* **14**, No. 9, 367–369. An attempt, knowing the maximum range on an inclined plane, to find the corresponding maximum on the horizontal.

Beevers, C. E. (1985) Motivating Mechanics. *Teaching Mathematics and its Applications* **4**, No. 2, 52–56. Uses computer programs to illustrate some practical problems in projectile and orbital motion.

Brancazio, P. J. (1985) Trajectory of a Flyball. *The Physics Teacher* **23**, No. 1, 20–23. Uses the principles of aerodynamics to describe the motion of a baseball in wind.

Brancazio, P. J. (1985) The Physics of Kicking a Football. *The Physics Teacher* **23**, 403–407. Illustrates some basic concepts of aerodynamics through their application to the trajectory of an American football.

Cooke, J. C. (1955) The Boundary Layer and 'Seam' Bowling. *The Mathematical Gazette* **39**, 196–199. Shows how the principles of laminar and turbulent flow can explain the swerve of a cricket ball.

Daish, C. B. (1972) *The Physics of Ball Games*. The English Universities Press, London. Outlines the basic physics of ball games, in a way which is scientific yet comprehensible to the average sportsman.

Davies, J. M. (1949) The Aerodynamics of Golf Balls. *Journal of Applied Physics* **20**, No. 9, 821–828. A series of experimental results on the lift and drag forces on a golf ball.

French, D. (1984) An Application to Archery. *The Mathematical Gazette* **68**, No. 445, 199–200. A short note to describe a simple application of projectile motion to archery.

Hart, D. (1987) Standardising the Shot. *Theta* **1**, No. 1, 3–5. A controversial suggestion that short shot putters should be compensated for their height.

Hart, D. & Croft, A. (1987) Some Thoughts on Projectiles. *Teaching Mathematics and its Applications* **6**, No. 2, 71–74. Application of the enveloping parabola to maximum-range problems.

Hart, D. & Croft, A. (1988) Is 20 mph too Fast? Safe Driving Speeds on Newly Surfaced Roads. *Mathematical Spectrum* (to appear). Discusses safe driving speeds and distances on resurfaced roads.

Haughland, O. A. (1983) A Puzzle in Elementary Ballistics. *The Physics Teacher* **21**, No. 4, 246–248. Describes why it is easier to miss a target when shooting a rifle up- or down-hill.

Hughes, D. E. (1985) Some Mathematics and Physics of Ball Games. *School Science Review* **67**, No. 238, 27–43. In the course of a long article the author describes the application of resisted and non-resisted projectile motion to tennis shots.

Lock, J. A. (1982) The Physics of Air Resistance. *The Physics Teacher* **20**, No. 3, 158–160. A short background article on the aerodynamics of a moving sphere (But see Mironer (1982) — Understanding Drag. *The Physics Teacher* **20**, 400, and Lock, The Author Adds Details, in the same issue).

Neie, Van E. Projectile Motion and Circular Orbits. *The Physics Teacher* **12**, No. 2, 102–103. Points out that the standard projectile results are based upon the constancy of the gravity vector.

Porter, G. J. (1981) New Angles on an Old Game. *American Mathematical Monthly* **88**, No. 4, 285–286. An interesting application to basketball.

Tait, P. G. (1893) On the Path of a Rotating Spherical Projectile. *Trans. Roy. Soc. Edinburgh* **37**, 427. An approximate analytical solution to the trajectory of a golf ball.

Tan, A. (1987) Shuttlecock Trajectories in Badminton. *Mathematical Spectrum* **19**, No. 2, 33–36. Uses intrinsic coordinates to model the flight of a shuttlecock.

Tan, A. (1987) Launch Angle for Running Cricket Ball Throw. *Theta* **1**, No. 2, 8–12. Discusses some aspects of projectile motion with a moving source.

Townend, M. S. (1984) *Mathematics of Sport*. Ellis Horwood, Chichester. Describes the application of mathematical techniques to a wide range of sporting activities from sailing to ice-skating.

Ward-Smith (1983) The Influence of Aerodynamic and Biomechanical Factors on Long Jump Performance. *Journal of Biomechanics* **16**, No. 8, 65–68. Uses the principles of aerodynamics and projectile motion to discuss long-jump techniques.

Williams, D. (1959) Drag Force on a Golf Ball in Flight and its Significance. *Quarterly Journal of Mechanics and Applied Mathematics* **12**, No. 3, 387–392. Discusses whether the long-hitting golfer has an unfair advantage over the ordinary golfer because of the reduced drag coefficient at very high speeds.

# References

Readers should also refer to the Select Annotated Bibliography in Appendix IV.

Alfonso, M. & Finn, E. J. (1970) *Physics*. Addison-Wesley, Reading MA.

Allen, C. W. (1963) *Astrophysical Quantities*. UOL, London.

Best & Bartlett (1987), Aerodynamic Characteristics of New Rules Javelin, Proc. Inst. Mech. Engineers, Conference Report, Leeds, Sept. 1987.

Bird, J. (1976) *Percy Grainger*. Elek, London.

Burghes, D. N., Huntley, I. & McDonald, J. (1982) *Applying Mathematics*. Ellis Horwood, Chichester.

Chorlton, F. (1978) *Textbook of Dynamics*. Ellis Horwood, Chichester.

Lichtenberg, D. B. & Wills, J. G. (1978) Maximising the Range of the Shot Put. *American Journal of Physics* **46**, No. 5.

Murphy, R. V. (1979) Maximum Range Problems in a Resisting Medium. *Mathematical Gazette* **63**, 423, 10–16.

Quadling, D. A. & Ramsey, A. R. D. (1966) *Elementary Mechanics Vol. II*. Bell, London.

Spiegel, M. R. (1974) *Advanced Calculus*. Schaum's Outline Series.

Wallace, J. M. & Hobbs, P. V. (1977) *Atmospheric Science, an Introductory Survey*. Academic Press, London.

The following books are also useful for the computer programming:

Oldknow, A. (1984) *Learning Mathematics with Micros*. Ellis Horwood, Chichester.

Oldknow, A. (1987) *Microcomputers in Geometry*. Ellis Horwood, Chichester.

# Index

## Mathematics and its Applications

*Series Editor:* G. M. BELL, Professor of Mathematics, King's College London (KQC), University of London

## Statistics and Operational Research

*Editor:* B. W. CONOLLY, Professor of Operational Research, Queen Mary College, University of London